Current Topics in Microbiology 145 and Immunology

Editors

R. W. Compans, Birmingham/Alabama · M. Cooper,
Birmingham/Alabama · H. Koprowski, Philadelphia
I. McConnell, Edinburgh · F. Melchers, Basel
V. Nussenzweig, New York · M. Oldstone,
La Jolla/California · S. Olsnes, Oslo · H. Saedler,
Cologne · P. K. Vogt, Los Angeles · H. Wagner, Ulm
I. Wilson, La Jolla/California

Molecular Mimicry

Cross-Reactivity Between Microbes and
Host Proteins as a Cause of Autoimmunity

Edited by M. B. A. Oldstone

With 28 Figures

Springer-Verlag
Berlin Heidelberg NewYork
London Paris Tokyo Hong Kong

MICHAEL B. A. OLDSTONE, M.D.

Dept. of Immunology, Scripps Clinic and Research Foundation
10666 N. Torrey Pines Road, La Jolla, CA 92037, USA

ISBN 3-540-50929-1 Springer-Verlag Berlin Heidelberg New York
ISBN 0-387-50929-1 Springer-Verlag New York Berlin Heidelberg

This work is subject to copyright. All rights are reserved, whether the whole or part of the material is concerned, specifically the rights of translation, reprinting, reuse of illustrations, recitation, broadcasting, reproduction on microfilms or in other ways, and storage in data banks. Duplication of this publication or parts thereof is only permitted under the provisions of the German Copyright Law of September 9, 1965, in its version of June 24, 1985, and a copyright fee must always be paid. Violations fall under the prosecution act of the German Copyright Law.

© Springer-Verlag Berlin Heidelberg 1989
Library of Congress Catalog Card Number 15-12910
Printed in Germany

The use of registered names, trademarks, etc. in this publication does not imply, even in the absence of a specific statement, that such names are exempt from the relevant protective laws and regulations and terefore free for general use.

Product Liability: The publisher can give no guarantee for information about drug dosage and application thereof contained on this book. In every individual case the respective user must check its accuracy by consulting other pharmaceutical literature.

Offsetprinting: Saladruck, Berlin; Bookbinding: B. Helm Berlin 2123/3020-543210 —
Printed on acid-free paper

Table of Contents

M. B. A. OLDSTONE: Overview: Infectious Agents as Etiologic Triggers of Autoimmune Disease 1

J. FROUDE, A. GIBOFSKY, D. R. BUSKIRK, A. KHANNA and J. B. ZABRISKIE: Cross-Reactivity Between Streptococcus and Human Tissue: A Model of Molecular Mimicry and Autoimmunity 5

W. VAN EDEN, E. J. M. HOGERVORST, E. J. HENSEN, R. VAN DER ZEE, J. D. A. VAN EMBDEN, and I. R. COHEN: A Cartilage-Mimicking T-Cell Epitope on a 65K Mycobacterial Heat-Shock Protein: Adjuvant Arthritis as a Model for Human Rheumatoid Arthritis 27

P. L. SCHWIMMBECK and M. B. A. OLDSTONE: Klebsiella pneumoniae and HLA B27-Associated Diseases of Reiter's Syndrome and Ankylosing Spondylitis 45

M. E. DIEPERINK and K. STEFANSSON: Molecular Mimicry and Microorganisms: A Role in the Pathogenesis of Myasthenia Gravis? 57

M. F. KAGNOFF: Celiac Disease: Adenovirus and Alpha Gliadin 67

G. B. TAKLE and L. HUDSON: Autoimmunity and Chagas' Disease 79

R. S. FUJINAMI: Immune Responses Against Myelin Basic Protein and/or Galactocerebroside Cross-React with Viruses: Implications for Demyelinating Disease . . . 93

R. T. DAMIAN: Molecular Mimicry: Parasite Evasion and Host Defense 101

T. DYRBERG: Molecular Mimicry and Diabetes 117

M. B. A. OLDSTONE: Molecular Mimicry as a Mechanism for the Cause and as a Probe Uncovering Etiologic Agent(s) of Autoimmune Disease 127

Subject Index 137

List of Contributors

You will find the addresses at the beginning of the respective contribution

Buskirk, D. R.
Cohen, I. R.
Damian, R. T.
Dieperink, M. E.
Dyrberg, T.
Froude, J.
Fujinami, R. S.
Gibofsky, A.
Hensen, E. J.
Hudson, L.
Hogervorst, E. J. M.

Kagnoff, M. F.
Khanna, A.
Oldstone, M. B. A.
Schwimmbeck, P. L.
Stefansson, K.
Takle, G. B.
Van der Zee, R.
Van Eden, W.
Van Embden, J. D. A.
Zabriskie, J. B.

Overview: Infectious Agents as Etiologic Triggers of Autoimmune Disease

M. B. A. Oldstone

Infectious agents, particularly viruses, are implicated in autoimmunity on the basis of three findings. First, autoimmune responses are made de novo or those already present are enhanced concomitant with infection by a wide variety of human DNA and RNA viruses. This point is strengthened by the second finding that, in experimental animals, both acute and persistent virus infections can induce, accelerate, or enhance autoimmune responses and cause autoimmune disease. For example, it has been shown that with the New Zealand mouse family, a genetically defined group of mice from which certain strains spontaneously develop autoimmunity, autoimmune manifestations normally present in NZB mice (DNA-specific antibodies, red blood cell-specific antibodies) or their (NZB × W) F_1 relatives (DNA-specific antibodies) are enormously enhanced by persistent infection with either a DNA (polyoma) or RNA (lymphocytic choriomeningitis, LCMV) virus; that is, antibodies form earlier and reach higher titers in the infected mice than in their uninfected counterparts (Tonietti et al. 1970; Lampert and Oldstone 1973). Further, NZW mice, which normally do not develop these autoimmune responses, do so upon polyoma or LCMV infections. Indeed, the responses in NZB, (NZB × W) F_1, or NZW mice are so marked that autoimmune diseases occur at a higher incidence with earlier time of death (NZB, (NZB × W) F_1] or appear de novo (NZW). These events were subsequently repeated with a number of viruses, including retroviruses (reviewed Oldstone 1972). Third, utilizing an investigative approach that focuses on one potential mechanism where by microbes cause autoimmunity, molecular mimicry, a number of etiologic agents have been defined as potential causes of autoimmune disease (Oldstone 1987).

Microbial agents can induce autoimmune responses by a variety of unique mechanisms, and several of these might occur during any one infection. For example, certain microbes have a mitogenic effect on unique lymphocyte subsets and hence can act as polyclonal activators. Because agents like mycoplasma can activate lymphocytes, and mycoplasma may contaminate bacterial and viral stocks, stringent evidence must be presented that the activation is directly due to the bacteria or virus and not mycoplasma contamination. Microbes can also direct the release of lymphokines and monokines. These molecules are important modulators of immune responses by acting as growth or differentiation factors or by regulating the expression of class I and class II major histocompatibility molecules. Hence, an infectious agent could release a lymphokine like interferon and induce the expression of class II molecules on cells

---

Department of Immunology, Scripps Clinic and Research Foundation, La Jolla, CA 92037, USA

like epitheloid cells in the thyroid or β cells in the islets of Langerhans of the pancreas. Such cells, which usually express limited or negligible amounts of MHC class II glycoproteins, may then present self-antigens (i.e., thyroglobulin, insulin receptor, etc.) in the context of induced class II molecules. The end result is hypothesized to be the priming of an immune response against self (BOTTAZZO et al. 1983). Additional microbes, especially viruses, can infect and selectively replicate in unique lymphocyte subsets (reviewed MCCHESNEY and OLDSTONE 1987). By their presence, activation, or replication, the microbe can cause immunosuppression or immunoenhancement. Finally, microbes can contain chemical structures that mimic normal host "self" proteins, an event termed molecular mimicry. Thus, an effector immune response, either B (humoral) or T (cytotoxic T cell), directed against the microbe might then also cross-react with "self" protein and induce autoimmunity. Several of the ways that infectious agents can trigger autoimmunity are:

- Polyclonal activator of B or T lymphocytes
- Enhanced expression of MHC class II molecules
- Enhanced expression of MHC class I molecules
- Bystander activation of immunocompetent cells through release of lymphokines or monokines
- Alteration of lymphocyte/macrophage function
  — Tropism for and lysis of unique cell subset(s)
  — Tropism for and alteration of differentiated function of unique cell subset(s) (no lysis)
- Molecular mimicry

This volume focuses on the evidence for or against molecular mimicry as a cause of autoimmunity. Major contributors in this area of research have provided chapters analyzing the evidence available. The group of investigators at the Rockefeller University have a long-term interest in the role of streptococcus in rheumatic heart diease. John Zabriskie reviews the story for streptococcal M protein, its shared homology and immunologic cross-reactivity primarily with myocardial tissue. The second chapter by W. van Eden relates a recent study implicating myobacteria, cartilage proteoglycans, and heat shock proteins with a cross-reactive T cell epitope in causing arthritis. Peter Schwimmbeck's chapter reviews the evidence suggesting that HLA B27 is an autoantigen that cross-reacts with *Klebsiella pneumoniae* and other bacteria. He argues that this may be the etiology and pathogenic mechanism for the nonrheumatoid arthritides, ankylosing spondylitis, and Reiter's syndrome. K. Stefansson explores the role of cross-reactive bacterial and viral antigens with the acetylcholine receptor as a possible cause of myasthenia gravis. Martin Kagnoff details his observation of the sequence and immunologic cross-reactivity of alpha gliadin and an adenovirus subtype as meaningful events in celiac disease. *Trypanosoma cruzi* causes the American form of trypanosomiasis called Chagas' disease. Drs. Hudson and Takle review the evidence supporting an autoimmune etiology for this parasitic infection. The next two chapters provide data suggesting a link for demyelinating disease and for diabetes with infectious agents and autoimmune response. Robert Fujinami has been interested in developing experimental animal models that help establish the principles of molecular mimicry (FUJINAMI and OLDSTONE 1985). In his contribution, Dr. FUJINAMI explores experimental and clinical findings relating a role for molecular mimicry in demyelinating and central nervous system disease. The chapter provided by Dr. DYRBERG is perhaps the most teasing. He provides preliminary data of a cross-reactivity between the insulin receptor, thyroid protein, and acetylcholine receptor with several infectious agents. His finding of a clinical condition associated with antibodies to the insulin receptor and of events in which a cross-reactivity between papillomavirus and the receptor occurred is most interesting.

The chapter by Dr. DAMIAN provides a balance. Reviewing the evidence for shared determinants between parasites and host cell antigens, Dr. DAMIAN argues for molecular mimicry as a basis of avoidance of immunologic recognition and cross-reactive immune response. Thus, in this view molecular mimicry may be beneficial and serve as a line of host defense.

The final chapter is positioned to review the overall perspective of molecular mimicry, how to use its principles in clinical investigation and list the conceptual traits by which autoimmune disease can occur.

Acknowledgements: This is publication number 5624-IMM from the Department of Immunology, Scripps Clinic and Research Foundation, La Jolla, CA 92037. This work was supported in part by USPHS grants AI-07007, NS-12428 and AG 04342.

References

Bottazzo GF, Pijal-Borrell PR, Hanafusa T (1983) Hypothesis: role of aberrant HLA-DR expression and antigen presentation in induction of endocrine autoimmunity. Lancet ii: 1115–1119

Fujinami, RS, and Oldstone, MBA (1985) Amino acid homology between the encephalitogenic site of myelin basic protein and virus: Mechanism for autoimmunity

Lampert PW, Oldstone MBA (1973) Host IgG and C3 deposits in the choroid plexus during spontaneous immune complex disease. Science 180: 408–410

McChesney MB, Oldstone MBA (1987) Viruses perturb lymphocyte functions: selected principles characterizing virus-induced immunosuppression. Ann Rev Immunol 5: 279–304

Oldstone MBA (1972) Virus induced autoimmune disease: viruses in the production and prevention of autoimmune disease. In: Membranes and viruses in immunopathology. Academic, New York, pp 469–475

Oldstone MBA (1987) Molecular mimicry and autoimmune disease. Cell 50: 819–820

Tonietti G, Oldstone MBA, Dixon FJ (1970) The effect of induced chronic viral infections on the immunologic diseases of New Zealand mice. J Exp Med 132: 89–109

Cross-Reactivity Between Streptococcus and Human Tissue: A Model of Molecular Mimicry and Autoimmunity

J. Froude[1], A. Gibofsky[2], D. R. Buskirk[1], A. Khanna[1], and J. B. Zabriskie[1]

1	Introduction	5
2	Hyaluronic Acid	7
2.1	Biochemical Homology	7
2.2	Serological Homology	7
2.3	Biological Homology	8
3	M Protein	8
3.1	Structural Homology	9
3.2	Serological Homology	10
3.3	Biological Significance	11
4	Carbohydrate	12
4.1	Structural Homology	12
4.2	Serological Homology	12
4.3	Biological Homology	13
5	Membrane Antigens	13
5.1	Structural Homology	13
5.2	Serological Homology	14
5.2.1	Heart	14
5.2.2	Brain	14
5.2.3	Kidney	15
5.2.4	Skin	15
5.3	Histocompatibility Antigens	16
6	Disease Associations	17
6.1	Rheumatic Fever	17
6.2	Post-Streptococcal Glomerulonephritis	18
7	Conclusions	20
References		22

1 Introduction

Structural homology between mammalian tissues and microbes has been demonstrated for over 50 years. Twenty years ago the term "molecular mimicry" was coined to describe antigenic structures common to host and protozoal parasites (Damian 1964). This term has since been expanded to include other microorganisms, and

---

[1] Laboratory of Bacteriology and Immunology, Rockefeller University, New York, NY, USA
[2] Multipurpose Arthritis Center, Hospital for Special Surgery, Cornell University Medical College, NY, USA

Table 1. A selection of cross-reactions between microbes and mammalian tissues

Organism	Tissue	Possible Disease Association	Ref.
Streptococcus pyogenes	Heart, brain, kidney, etc.	Rheumatic fever	ZABRISKIE 1985
Streptococcus mutans	Heart	?	VAN DE RIJN et al. 1976
Klebsiella	HLA-BW27 lymphocytes	Ankylosing spondylitis Reiter's syndrome	EBRINGER 1978; SEAGER et al. 1979; SCHWIMMBECK et al. 1987
Salmonella	Mouse tissues	Salmonella infection	ROWLEY and JENKINS 1962
Pneumococci 50% gram-negative bacteria	Blood group substances	? Susceptibility to infection	FINLAND and CURNEN 1940; SPRINGER et al. 1961
Escherichia coli	Colon tissue	Ulcerative colitis	PERLMAN et al. 1965
Tubercle bacillus	Cartilage proteoglycanase	Rheumatoid arthritis	VAN EDEN et al. 1985
Trypanosoma cruzi	High density lipoprotein	Chagas' disease	PRIOLI et al. 1987
Coxsackie B virus	Cardiac myosin	Cardiomyopathy	ROSE et al. 1986

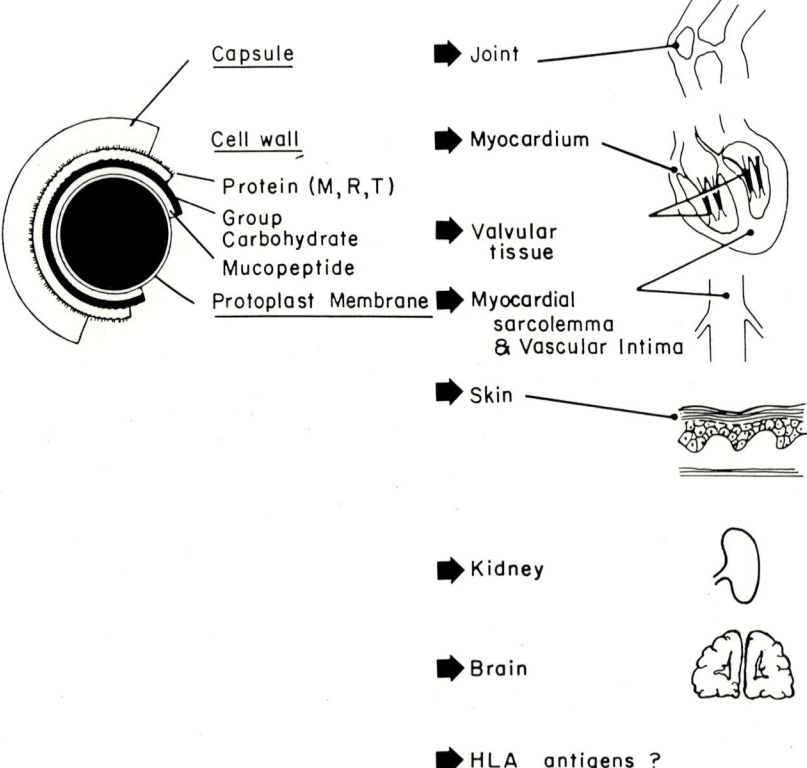

Fig. 1. Schematic representation of the various structures of the group A streptococcus. Note the wide variety of cross-reactions between its antigens and mammalian tissues

knowledge of the phenomenon has increased greatly with the application of newer immunochemical and molecular biological techniques. Cross-reacting antibodies to the shared antigens may arise, and a partial list of these and the disease states with which they may be associated is shown in Table 1.

The question that remain to be fully answered are: How are these cross-reactions important in the pathogenesis of disease? What is the biological nature of immunological cross-reaction? Is there a genetic predisposition to the ill effects of immunological cross-reaction, and what is its precise nature?

The focus of this review will be on antigenic similarities between the group A streptococcus and its natural host, man, and their relationship to acute streptococcal infections and their nonsuppurative sequelae.

Figure 1 is a diagram of the intact streptococcus. The complex structure would seem to afford considerable potential for cross-reaction with the host. While examining the evidence, it is always important to bear in mind that the streptococcus, like many other organisms, has Fc binding proteins that may lead to nonspecific binding (HOLM et al. 1978). The relevant biochemical components of these microbe-host reactions will now be discussed in detail.

2 Hyaluronic Acid

2.1 Biochemical Homology

First isolated from bovine vitreous humor (MEYER and PALMER 1936), hyaluronic acid was shown to contain equimolar concentrations of N-acetyl glucosamine and glucuronic acid (MEYER 1958). Soon after, it was isolated in pure form from the capsules of group A streptococci and shown to be structurally indentical to the mammalian moiety (KENDALL et al. 1937). Although there is some controversy concerning the linkage of hyaluronic acid to protein moieties in mammalian tissues (SANDSON and HAMMERMAN 1962), there is no evidence that streptococcal hyaluronic acid binds to other proteins, since the capsule can be isolated in pure form.

2.2 Serological Homology

Several investigators were unable to raise heteroantibodies to streptococcal hyaluronic acid (SEASTONE 1939; HUMPHREYS 1943; QUINN and SINGH 1957). This was attributed to absolute structural identity with the hyaluronic acid of the animal in which the antibody was to be raised. FILLET et al. (1986) showed, however, that high titers of *nonprecipiating* antibodies to streptococcal hyaluronic acid could be induced in rabbits, indicating that heteroantibodies could be demonstrated under the appropriate test conditions.

UNDERHILL (1982) noted that a number of animal sera contained naturally occurring antibodies to hyaluronic acid, and this was later shown in human sera also (FAARBER et al. 1984). In terms of human disease, FILLEIT et al. (1984) noted that sera from patients with post-streptococcal nephritis contained antibodies that cross-reacted with gomerular proteoglycans. These antibodies were partially inhibited by hyaluronate, suggesting that at least some of the antibodies might have been directed against the hyaluronic acid capsule of the inciting streptococcal strain.

2.3 Biological Homology

SEASTONE (1939) showed that the presence of an *encapsulated* group C streptococcus was important in the infection of guinea pigs. In group A streptococcal infections of other experimental animals, however, the results were not as clear. Although it is true that prior treatment with hyaluronidase could protect mice from up to 10 minimal lethal doses (MLD) of group A streptococci, pretreatment with antisera to M protein protected them from up to 100 000 MLD of group A organisms, demonstrating the greater virulence potential of the M protein (ROTHNARD 1948).

HIRSCH and CHURCH (1960) showed the presence of a "factor" in human sera that promoted phagocytosis by human leukocytes and rapid killing of encapsulated group A streptococci under in vitro conditions. Under the same conditions, rabbit leukocytes did not do this. The "factor" was not antibody and was thermolabile. STOLLERMAN et al. (1963) confirmed these observations and also demonstrated that certain human sera were deficient in the "factor." The "factor" may be a component of complement, but efforts to date to identify this molecule have been unsuccessful.

An organism capable of wearing an "overcoat" structurally identical to molecules present on numerous mammalian tissues should have a distinct advantage in gaining entrance into the host. This has been difficult to demonstrate in humans, however. Perhaps Fillit's discovery of nonprecipitating antibodies to streptococcal hyaluronic acid may prove the most fruitful avenue of research into the possibility that these antibodies play a causal role in certain diseases.

3 M Protein

Decades of elegant experiments by LANCEFIELD (1962) have established that the M protein is an important, if not the most important, virulence factor in group A streptococcal infections of humans. Her serological studies revealed that there are at least 80 serologically different M proteins, all antiphagocytic in vitro.

3.1 Structural Homology

Fischetti and coworkers first delineated M protein as an alpha-helical coiled-coil structure bearing a close resemblance to other coiled-coil proteins (HOSEIN et al. 1979; MANJULA and FISCHETTI 1980; PHILLIPS et al. 1981; FISCHETTI et al. (1988) (Fig. 2).

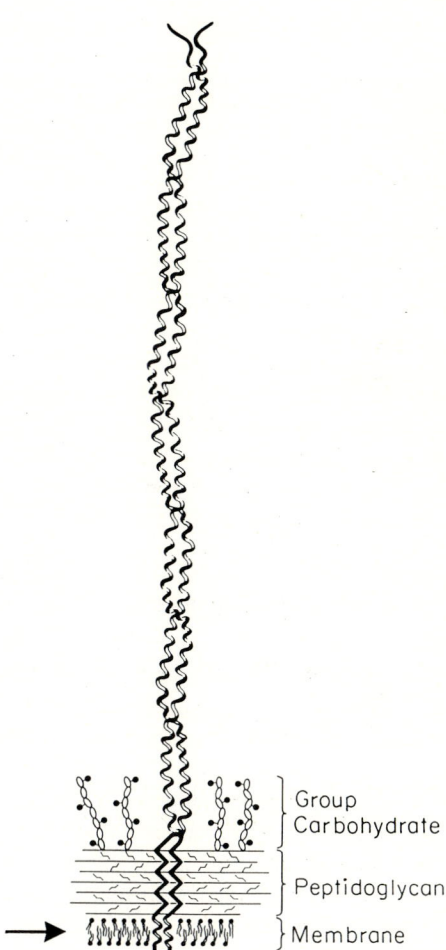

Fig. 2. Schematic drawing of the two-dimensional structure of the M protein moiety and its relationship to the streptococcal cell. *Arrow* denotes that portion of the M protein which is anchored to the membrane

As shown in Table 2, all three M proteins for which the entire amino acid (AA) sequence is known bear close homology to myosin and tropomyosin, also coiled-coil proteins. Keratin also has close AA sequence similarity to M proteins, in many cases as closely related as the cardiac muscle proteins. These findings have been confirmed and extended by other investigators (DALE and BEACHEY 1982, 1986a).

Despite a good deal of linear AA sequence homology between these coiled-coil structures, it is more likely that secondary and tertiary structural epitopes have greater functional significance than linear AA homology alone. For instance, internal

Table 2. Sequence homology of M proteins with mammalian proteins

M protein type	Mammalian protein	% Identity
M24	Myosin heavy chain (rabbit cardiac)	24
	Tropomyosin (human cardiac)	23
	Keratin (type I, bovine)	22
	Lamin (human)	18
M5	Tropomyosin (human fibroblast)	27
	Myosin heavy chain (rabbit cardiac)	24
	Keratin (type I, bovine)	22
	Tropomyosin (human cardiac)	20
M6	Myosin (rat cardiac)	23
	Keratin (type I, bovine)	19
	Tropomyosin (human cardiac)	19
M12	Keratin (type I, human)	21
	Myosin heavy chain (rabbit cardiac)	20

AA are more conserved in M protein than external AA, which are more conformationally dependent and variable. These external AA may be more important in the observed cross-reactions between M proteins and mammalian tissues, a subject of current research in Dr. Fischett's laboratory.

SARGENT et al. (1987) showed that AA region residues 164–197 of highly purified peptic extracts of M protein 5 (pepM5) cross-reacted with a sarcolemmal protein of approximately 40K; in contrast, synthetic M5 residues cross-reacted primarily with myosin (DALE and BEACHEY 1986a).

3.2 Serological Homology

KAPLAN (1963) was the first to show that immunization of rabbits with streptococcal cell walls could elicit antibodies cross-reactive with cardiac tissue. These antibodies could be absorbed *only* with certain M protein-specific strains of group A streptococci.

Once the AA sequence of a number of M proteins became known, it became possible to localize those particular areas of the M protein that were cross-reactive with mammalian tissues. A purified preparation of M5 protein (pepM5) not only cross-reacted with human arcolemma antigens but was also part of the M protein moiety involved in the opsonic reaction (DALE and BEACHEY 1982). Then it was shown that certain M types (M6, M19), but not others (M24), also contained these cross-reactive epitopes and were also opsonic. Of interest was the finding that M5 contained three cross-reactive epitopes, whereas M6 and M19 had only two (DALE and BEACHEY 1985).

Antisera to M proteins stained several cardiac antigens of differing molecular weights. Examination of these sera with cardiac myosin showed cross-reaction with an M protein moiety. However, some sarcolemmal antigens detected by the M protein antisera did not, indicating epitope homology *unrelated* to myosin.

SARGENT and coworkers (1987) produced a series of synthetic peptides spanning the entire pepM5 molecule. AA residues 164–197 formed the only area to cross-react with a 40K protein of heart. This area is related to the opsonic region of the molecule and is shared by several heterologous M proteins. In contrast, AA residues 84–116 contain the residues cross-reactive with myosin (DALE and BEACHEY 1986a). It should be pointed out, however, that at the N-terminal of some M proteins are moieties that elicit protective opsonic antibodies without tissue cross-reaction, and this could be important in the development of antistreptococcal vaccines (DALE and BEACHY 1986b).

The promising implication of this distinction between protective and cross-reactive epitopes has been countered to some extent by the latest studies of BRONZE et al. (1988) demonstrating that, at least for M19, the protective and cross-reacting epitopes are *both* localized within the first 24 N-terminal AA of the molecule.

Turning attention away from the N-terminal, BESSEN and FISCHETTI (1988) have noted that many M proteins have a constant region extending from the pepsin cleavage site to the anchor region of the membrane. Mice immunized intranasally with synthetic peptides of this region were protected against intranasal challenge with the homologous organism, providing evidence for protection *without* involving the N-terminal structure of the M protein. Whether these protective epitopes also contain cross-reactive epitopes is currently under investigation.

Both the diversity and extent of the cross-reactions between the M protein and human tissues are emphasized by GORONCY-BEARNS et al. (1987), who found that a monoclonal antibody against human glomeruli cross-reacted with pepM6 and pepM12 only. The cross-reactive glomerular protein was noted to be a 40K protein. Using synthetic peptides representing residues 1–26 of M1 protein yielded a reaction with glomerular tissues. No reaction occurred if residues 20–22 were omitted (KRAUS and BEACHEY 1988). By using differing synthetic peptides, the glomerular cross-reactive epitope was shown to have an Ile-Arg-Leu-Arg sequence at position 23–26.

3.3 Biological Significance

A group A streptococcal protein whose amino acid sequence cross-reacts with at least two mammalian cardiac proteins might be thought to have considerable relevance to the pathogenesis of rheumatic fever (RF). Yet evidence implicating these cross-reactions is surprisingly scant. Antibodies to myosin have been detected in the sera of patients with acute rheumatic fever (ARF) and are also present in a high percentage of sera from patients with uncomplicated streptococcal infections (CUNNINGHAM et al. 1988). Further, myosin is primarily an *internal* protein of cardiac muscle cells and may not be easily exposed to M protein cross-reacting antibodies.

Perhaps the more interesting cross-reactive epitope in terms of disease pathogenesis is the M protein peptide 164–197, which recognizes a *sarcolemmal* protein of 40K. Our own studies of sera from individuals with ARF also identify a cardiac sarcolemmal protein of approximately the same molecular weight (unpublished data). These heart reactive sera recognize a streptococcal membrane protein of 30K that absorbs *all* of the heart reactive binding of ARF sera to cardiac antigens (VAN DE RIJN et al. 1977) and may represent the most significant cross-reaction between tissue and microbe.

The possibility that myosin might play a direct role in induction of myocarditis comes mainly from the work of Rose and his colleagues (ROSE et al. 1986; NEU et al. 1987a) in experimental coxsackie virus B_3 infections of mice. Following the initial infection with the virus, an immune-mediated second phase occurs that results in chronic progressive myocarditis. Heart reactive antibodies are present in high titers in sera of the susceptible murine strain but not in the resistant strain. Immunoglobulins deposit in the tissues of these mice, and these heart reactive antibodies are specifically absorbed with the *cardiac* isoform of myosin. More recently, these investigators were able to induce all the signs and symptoms of virally induced myocarditis by injecting *only* cardiac myosin into the susceptible strains (ROSE et al. 1987). Of interest was their observation that antibodies to the virus itself do not cross-react with myosin (NEU et al. 1987b), suggesting that release of myosin per se initiates the autoiimune process in the mice and is also capable of inducing the disease even in naive animals of the susceptible strain.

A somewhat different view is presented by the data of SAEGUSA et al. (1986), who prepared a number of monoclonal antibodies against coxsackievirus B_4, which were tested by immunofluorescence against a variety of mammalian tissues, including the heart. One of these monoclonal antibodies exhibited neutralizing titers against the virus and also bound to rabbit and mouse heart tissues, but *not* to human heart tissue. Thus, these biologically important cross-reactions between coxsackieviruses and mammalian tissue may depend not only on the strain of virus tested (i.e., B_3 versus B_4), but also may be quite specific for a component of the myosin of a given mammalian species.

4 Carbohydrate

4.1 Structural Homology

MCCARTY (1970) and co-workers identified the structure of group A carbohydrate as a polysaccharide chain consisting of approximately nine repeating units of rhamnose capped by an *N*-acetyl glucosamine chain. This chain is immunodominant, giving group specificity to group A streptococci. The polysaccharide is present in many mammalian tissues.

4.2 Serological Homology

GOLDSTEIN et al. (1968) first reported an antigenic similarity between glycoproteins extracted from human and bovine heart valves and streptococcal group A carbohydrate. This similarity between the two antigens rested on the fact that both contained the immunodominant sugar, *N*-acetyl glucosamine, since antisera prepared to the valvular glycoprotein also bound to the group A streptococcus and was absorbable both by glycoprotein and group A carbohydrate. The fact that this valvular glycoprotein material was extracted following trichloracetic acid and 8 *M* urea treatment of the valves, coupled with its isolation from skin, cornea, and cartilage, suggests in the light of the recent evidence (see Sect. 2) from FILLIT et al. (1984)

that the material in question might be a proteoglycan. This suggestion is strengthened by the data of GOLDSTEIN and CARAVANO (1967), who observed that sera from patients with RF reacted to this glycoprotein. Fillit (personal communication) has observed the same reactivity to proteoglycans in the sera of such patients. Alternatively, the reaction could be secondary to N-acetyl glucosamine present in the hyaluronic acid moiety in group A streptococci and mammalian tissues.

GOLDSTEIN demonstrated that the cross-reactivities of group A streptococcal carbohydrate and valve glycoprotein were reciprocal. For example, by both immunodiffusion and immunofluorescence, reactivity of antisera to group A streptococci was 100% absorbed by group A carbohydrate and 50% absorbed by the glycoprotein. Conversely, the activity of antisera to the glycoprotein was completely abolished by the urea-extracted material and partially absorbed by the carbohydrate, suggesting that both antisera shared antibodies with a common immunodeterminant yet retained a 50% specificity for the immunizing antigen.

4.3 Biological Homology

Do antibodies to group A carbohydrate found in the sera of patients with RF contribute to valvular damage? DUDDING and AYOUB (1968) approached this question by examining the sera of patients with and without rheumatic heart diesease for the presence of antibodies binding to radiolabeled group A carbohydrate. Both groups showed high titers at the onset of disease. During the subsequent 5 years, however, the patients with heart disease continued to have high titers, while the titers of those without declined dramatically. FUJIKAWA et al. (1987) have confirmed this finding at the onset of RF for Korean patients, but long-term follow up was not a part of their study.

5 Membrane Antigens

5.1 Structural Homology

In terms of microbe-host cross-reactions, perhaps the most extensively studied streptococcal microbial structure has been its membrane. FREIMER (1963) found that it contains 71% proteins, 26% lipids, and 3% sugars. Two-dimensional gel analysis shows 300 distinct proteins (VAN DE RIJN and TRISCOTT 1984). PANCHOLI an FISCHETTI (1987) noted that M protein extends further into the cell wall than previously thought and is anchored in the cell membrane. This is important since cross-reactions attributed to membrane antigens may, in fact, be from M protein COOH-terminal epitopes.

In spite of the numerous serological observations to be described below, we still have little structural data to account for these cross-reactions with so many mammalian tissues. It is to be hoped that future studies will concentrate on the molecular basic of these cross-reactions in much the same manner as described for the M protein moieties.

5.2 Serological Homology

5.2.1 Heart

ZABRISKIE and FREIMER (1966) immunized rabbits with streptococcal membranes uncontaminated by the cell wall. This elicited antibodies that cross-reacted with human cardiac, skeletal and smooth muscle of blood vessels (but not uterine smooth muscle) and did not distinguish between rheumatic and nonrheumatic heart tissue. Sera from patients with ARF stained similarly (ZABRISKIE et al. 1970). The titers were highest during the acute disease, then declined over a 2–3-year period. The heart staining both in the rabbits and humans could be abolished by absorption with pure but undifferentiated streptococcal membrane antigens. Using a series of purification steps with this membrane material, VAN DE RIJN et al. (1977) found that four proteins with a molecular weight range of 22K–32K were capable of absorbing *all* heart reactive antibodies in ARF patients. An interesting observation in these studies was that rabbits immunized with membranes produced antibodies that did not block the heart reactive antibodies of sera from patients with RF. The converse was also true, suggesting that the rabbit antibody and the human antibody reacted with separate epitopes in the heart tissue. This may go some way towards explaining the curious fact that there is no satisfactory animal model for RF, in spite of numerous attempts to find one.

After the earlier observations carried out with polyclonal sera, CUNNINGHAM et al. (1984, 1986) began to explore these cross-reactions in more precise detail using monoclonal antibodies and mice immunized with isolated streptococcal membranes. The hybridoma clones were then screened against streptococci and cardiac tissue. They showed that streptococcal membrane antigens induced antibodies that bound to the *C*-terminus of the pepM5 protein and also the heavy chain portion of myosin (KRISHER and CUNNINGHAM 1985). More recently, CUNNINGHAM et al. (1986) found that some of the monoclonals reacted with the meromyosin light chain. Surprisingly, these antibodies also bound keratin, other alpha-helical coiled-coil proteins, and DNA (CUNNINGHAM and SWERLICK 1986). This strongly suggests that configurational and/or charge properties common to a number of mammalian antigens are what is being recognized. In an attempt to elucidate the molecular biology of this, we are currently screening a lambda gT-11 gene bank with monoclonal antibodies against membranes in order to explore further the streptococcal antigenis cross-reaction with human sarcolemma.

5.2.2 Brain

KINGSTON and GLYNN (1971) were the first to observe that antisera to group A streptococci stained astrocytes of the brain. This interesting reaction between brain and streptococci was reviewed by HUSBY et al. (1976), who noted that the majority of sera from patients with Sydenham's chorea specifically stained the cytoplasms of subthalamic and caudate nuclei. Of sera from patients with RF and *without* chorea, 25% also stained these nuclei, suggesting that more than just the presence of these antibodies was required to initiate chorea.

The streptococcal association with chorea was firmly established when these antibodies could be completely absorbed by streptococcal membrane antigens and were found specific for group A streptococcal structures.

5.2.3 Kidney

The initial observations were based on early work by MARKOWITZ and coworkers (1960), who noted that nephritogenic streptococcal strains, when injected into implanted intraperitoneal diffusible chambers, produced focal glomerular nephritis in rats. Suspecting that there might be antigenic similarities between streptococci and renal glomeruli antigens, Markowitz isolated cell walls and membranes from streptococci, prepared antisera to these fractions (MARKOWITZ and LANGE 1962, 1964), and found that only streptococcal membranes contained antigens that cross-reacted with human glomeruli. The cross-reactivity was also primarily found in those strains capable of causing nephritis. HOLM (1967) expanded and confirmed these findings with the exception that he demonstrated several distinct cross-reactions between streptococcal membranes and human glomerular extracts; these were not specific for nephritogenic streptococcal membranes alone.

Subsequently, RAPAPORT et al. (1969) showed that antisera prepared against purified streptococcal membranes and injected into animals bound to renal basement membrane antigens. In an effort to be more precise, FITZSIMMONS et al. (1987) recently injected mice with streptococcal membranes and isolated one monoclonal antibody that bound specifically to renal basement membrane antigens.

5.2.4 Skin

With the classic work of RAPAPORT and CHASE (1964) came evidence that streptococcal antigens also shared antigenic determinants with atigens of skin and with transplantation antigens. They showed that animals primarily sensitized with heat-killed group A streptococci and later purified group A membrane antigens had an accelerated rejection of skin homografts compared with unimmunized controls. The specificity of the reaction was attested to by the fact that immunizations with other groups of streptococci as well as other gram-positive organisms were ineffective in causing the accelerated rejection.

Definition of the exact nature of the tissue antigens involved and the pathogenic mechanisms responsible for the homograft rejection lay dormant for almost 2 decades. While RAPAPORT and CHASE (1964) favored the concept that homograft rejection was related to epitopes present on streptococci and tissue transplantation antigens, the recent work of CUNNINGHAM et al. (1986) suggested that the rejection could be related to streptococcal cross-reactions with keratin. An alternative candidate might be proteoglycan, a constituent of all tissue membranes, since FILLIT et al. (1984) have shown that the sera of patients with post-streptococcal sequelae do contain antibodies to proteoglycans.

5.3 Histocompatibility Antigens

HIRATA and TERASAKI (1970) showed that allogenic antisera against human lymphocytes were specifically inhibited by M protein from group A streptococcus type 1. Analogous M proteins from other streptococcal types had little or no inhibitory activity. In order to exclude nonspecific inhibition of cytotoxicity due to complement activation, the soluble test substance was incubated with complement for 1 hour before addition of the lymphocytes, and about a 20% decrease in cytotoxicity was found at lower concentrations of antiserum. They concluded that this was evidence of inhibition of human allogenic serum by a bacterial protein that could well have a structure common to human histocompatibility antigens. They pointed out that strong cross-reactivity could lead to breakdown of tolerance to autologous transplantation antigens. Why M1 protein should be the most effective inhibitor of cytotoxicity was not clear.

In contrast, TAUBER et al. (1976a, 1980) argued that inhibition of cytotoxicity by nonspecific activation of complement would have been greater had the conditions of the assay varied (i.e., the introduction of a solid phase secondary to the introduction of lymphocytes). Moreover, immune complexes formed when M protein bound to nonspecific antibodies might have consumed complement. These authors went on to study human and rabbit hyperimmune sera directed against different structures from various streptococcal strains and found that none exerted specific cytotoxicity for any given HLA type.

Thus, the specificity of the cross-reactivity reported by Hirata and Terasaki remains in doubt, particularly since the M protein preparations used in 1970 in their studies contained numerous other streptococcal antigens unrelated to M protein.

Two additional pieces of information suggest antigens shared between the streptococcus and the MHC. The first is a report by ROSES et al. (1973) in which a human skin graft incompatible with the recipient for HLA-B8 showed a hyperacute "white graft" rejection. Skin allografts from a second donor who was phenotypically compatible and a third who was incompatible for antigen W28 were rejected in a nonaccelerated fashion. The recipient had had a group A streptococcal and staphylococcal infection in a hip ulcer 3 months prior to the skin graft, and capillary migration studies of his mononuclear cells showed increased cellular sensitivity to streptococcal membrane antigens compared with controls.

The second piece of information comes from DOSSETOR et al. (1978), who observed that a kidney transplant recipient rejected a graft from her mother. Although there was HLA incompatability for the A3, B7 antigens, cross-match detected no humoral sensitization against her mother of cells of a panel of lymphocytes except for a weak ADCC reaction to the mother. One month before transplantation, she developed and was treated for *Streptococcus faecalis* peritonitis. Re-examination of her sera taken just before transplant surgery now showed cytotoxic antibodies to a wide panel of HLA alleles and to, most importantly, HLA antigens of her mother. Absorption with membranes from the offending group D streptococcus abolished the cytotoxicity, whereas cell walls did not. Given that these are only case reports, they do suggest that bacterial infection can stimulate the rejection of transplanted tissue that might otherwise have been accepted.

We have described primarily the evidence for rejection of transplanted tissues secondary to micro-transplantation antigen cross-reactions; however, when administered in the correct dosage, these microbe antigens can also induce tolerance to tissue antigens. ELLIS et al. (1976) clearly showed that mice, previously immunized with 100 µg of streptococcal membrane, acutely rejected a heterologous embryonic heart transplanted in to the ear of the immunized recipient. However, when the mouse was given a tolerizing dose of 500 µg of membrane, the organ transplant survived for the duration of the experiment (45 days). The implications of these studies for future research in human transplants warrant further exploration.

6 Disease Associations

Clinical details of diseases in which streptococcus — host interactions are thought to play a role have been described in several reviews and will only be briefly touched on here (KAPLAN and FRENGLEY 1969; ZABRISKIE 1985).

We postulate that three conditions of microbe-host molecular mimicry lead to disease. First, cross-reaction between the microbe and the host; second, a unique humoral/cellular immune response of the host to self-antigens that are shared with the microbe; third, a genetic predisposition to disease in the host.

6.1 Rheumatic Fever

That group A streptococci contain antigens that cross-react with tissue in the heart, joints, skin, and caudate nuclei has been amply demonstrated. Sera from patients with ARF, moreover, contain these cross-reactive antibodies. They are present at the time of the disease, tend to disappear during quiescent periods, and reappear at the time of recurrence.

The pathologic lesions of RF suggest a complex interaction of humoral and cellular autoimmune mechanisms. READ et al. (1974) and later REID et al. (1980) showed an increased response to streptococcal antigens, particularly membrane antigens, which persisted for 2 years after the initial attack. GOWRISHANKAR and AGARWAL (1980) also noted that mononuclear cell populations from patients with ARF reacted with cardiac tissue. Isolation and expansion of T cells from valve specimens obtained from patients with active or inactive RF suggest that they respond to membrane antigens, not carbohydrate or M protein moieties (unpublished data). In contrast, DALE and BEACHEY (1987) showed that pepM5 activates normal donor peripheral blood lymphocytes and that they are cytotoxic against cultured human cells including heart cells, but not animal cells. Rabbit antisera to pepM5 partially inhibit cytotoxicity against heart cells but have no effect on other target cells. This suggests that some of the effector lymphocytes are recognizing M protein cross-reactive antigens.

That heredity plays a part in susceptibility to RF has been suggested by generations of researchers since CHEADLE (1889), whose wife and son suffered from the disease, stated that the genetic factors predisposing to RF were underestimated. The tendency of RF to affect more than one member of a family, its recurrence in the same individual, its relative rareness compared to with infection, and its association with an altered immune response have lead to the search for patterns of inheritance. WILSON et al. (1943) suggested that susceptibility to RF might result from the inheritance of a single gene in an autosomal recessive fashion. In contrast, TARANTA et al. (1959), after studying twins, postulated that the mode of inheritance might be via a dominant gene with limited penetrance. Monozygotic twins have a higher concordance for the disease than dizygotic twins or families, but this was not greater than that for tuberculosis or poliomyelitis, although there is a marked concordance for the individual clinical manifestations of RF (DISCIASCIO and TARANTA 1980). A similar pattern of unexpectedly low concordance in identical twins for a number of autoimmune diseases has been reported (WILLIAMS et al. 1980; BARNETT et al. 1981).

AYOUB et al. (1986) and ANASTASIOU-NANA et al. (1986) found a significantly higher incidence of HLA-DR4 in white American patients with RF and HLA-DR2 in black American patients. In contrast, HAFEZ et al. (1985) studying multiplex families with RF in Egypt found an increased association between the disease and HLA-B5. JHINGAN et al. (1986) noted increased HLA-DR3 for patients in Northern India. In none of these studies was the relative risk very high. The observed differences could result from differences in ethnic background or clinical selection of the study groups. A genetic tendency associated with the MHC is not negated, however, as dicussed in greater detail below.

In contrast to the HLA associations noted above, PATARROYO et al. (1979) in conjunction with our group found a B-cell alloantigen present in 75% of all RF patents tested as compared with 10% of the normal population. This marker was unrelated to any HLA antigen so far tested. By preparing monoclonal antibodies against the B cells of patients with RF, we found one, D8/17, that defines an antigen of considerable biological interest. In 100% of RF patients so far tested (see Table 3), the antigen is present on at least 20% of B cells. In normal people or patients with acute post-streptococcal glomerulonephritis the percentage of antigen-positive B cells ranges from 0%–5%. This marker, in common with Patarroyo's original alloantigen, is unrelated to any known HLA specificity. The patients come from widely differing geographical and ethnic backgrounds, and similar results have been found in India, New Zealand, and Mexico (unpublished data). What is not known and currently under investigation in our laboratory is whether the marker is present *prior* to onset of disease (i.e., predictive) or whether it appears in an expanded population of B cells *after* the disease. If it is predictive, it has exciting implications for the detection of susceptible individuals.

6.2 Post-Streptococcal Glomerulonephritis

Similarities between streptococcal and glomerular antigens led to a series of experiments to test the biological activity of this molecular mimicry. MARKOWITZ (1969) was able to induce glomerulonephritis in monkeys following their immunization with streptococcal membrane fractions known to cross-react with human glomeruli. Examination of the kidneys in these animals revealed that the pathological findings bore a close resemblance to those in human post-streptococcal glomerulonephritis.

Using antisera prepared against streptococcal membranes in sheep, a similar disease was induced in dogs (RAPAPORT et al. 1969).

In the human disease a number of studies have implicated kidney-specific antibodies in glomerulonephritis (LANGE et al. 1948; LUI and MCCRORY 1958), but the exact nature of the antigens and antibodies remained in doubt until recently. FILLIT et al. (1984) has re-examined the question of the nature of these antibodies and found that a number of patients with both acute and chronic post-streptococcal glomerulonephritis reacted to renal basement membrane proteoglycans and that the immunodominant determinant was the heparan sulfate moiety of the proteoglycan.

The recent evidence that M proteins share antigenic determinants with renal glomerular antigens coupled with observations that the cross-reaction is limited to the Ile-Arg-Leu-Arg tetrapeptide of the M1 protein raise another possibility; namely, that M proteins might also be involved in either the acute or chronic forms of post-streptococcal glomerulonephritis. It remains to be seen whether the sera of either group of patients contain antibodies that bind to this tetrapeptide, and whether these antibodies are associated with the clinical course of the disease.

Several studies have suggested an important role for cellular immunity in the initiation and/or progression of disease. JONES (1951) recognized that mononuclear cells were present in the glomeruli of patients dying of acute post-streptococcal glomerulonephritis. Since that time, numerous investigators (reviewed in FILLIT and ZABRISKIE 1982) have demonstrated that mononuclear cells play an important role in inducing various forms of glomerulonephritis and that macrophages are essential to the induction of disease. The recognition that the specific target of these cells in the glomeruli is renal tissue antigens was noted first by BENDIXIN (1968) and later ROCKLIN et al. (1970), when they demonstrated that mononuclear cell populations of patients with progressive glomerulonephritis reacted to renal glomerular antigens. These studies were expanded by FILLIT et al. (1978). He showed that patients with progressive glomerulonephritis reacted primarily to glycosidase-treated glomerular membrane extracts but *not* to native or collagenase-treated membranes, and that this reactivity was not related to the level of circulating immune complexes or to autoantibodies to glomerular basement membranes. The importance of cross-reactions to streptococcal antigens and renal antigens in human glomerulonephritis is strengthened by the finding that patients with progressive glomerulonephritis have an increased cellular reactivity to streptococcal antigens (ZABRISKIE 1971) compared with controls and nonglomerular renal disease and the observation by FILLIT et al. (1978) that the majority of patients who react to glycosidase-treated glomerular basement membrane also react equally well to streptococcal membrane antigens. Further examination of the cell types involved in these mononuclear cell populations in patients with streptococcal glomerulonephritis indicated that both T-helper and T-suppressor cells were depressed but that the number of streptococcal membrane antigen binding cells had increased (WILLIAMS et al. 1981).

Is there a genetic predisposition to acute post-streptococcal glomerulonephritis? When LAYRISSE et al. (1983) examined 18 families of index cases with unsubclassified forms of the desease, the data obtained were neither sufficient to reject the hypothesis of linkage with HLA nor to support it. In contrast, SARAZUKI et al. (1979) found a significant increase of HLA-D En (relative risk 9.0) in patients with acute post-streptococcal glomerulonephritis in Japan. Most interestingly, FORSBERY and JOHNSON

(1987) noticed a relationship between men aged 16–25 with primary glomerulonephritis progressing to chronic renal failure and HLA-B40. These patients had an associated increase in antistreptolysin 0 titer. In females or males over 25 this association was not apparent.

7 Conclusions

As has been discussed above, biological mimicry is common and group A streptococcus provides a striking example, sharing many antigens with its human host. Do cross-reacting antibodies to these antigens play a part in the pathogenesis of illnesses like RF, in which autoimmune disease follows infection?

Autoantibodies are commonly found in healthy individuals and their harmlessness may be due to their low titer in this group. Heart reactive antibodies, for instance, are found in low titers in some patients after uncomplicated streptococcal pharyngitis but appear to be of no clinical importance. In contrast, RF patients have higher titers that last longer and return with recurrence of disease (ZABRISKIE et al. 1970).

It is also known that autoantibodies may be a harmless epiphenomena of tissue damage. Heart-specific antibodies have been detected following both open heart surgery and acute myocardial infarction; as such their appearance in serum is probably a response to cellular antigens released into the circulation. In RF, however, heart reactive antibodies are absorbed not only by cardiac antigens but also by a solubilized and highly purified fraction of group A streptococcal membranes, suggesting a causal relationship to a streptococcal cross-reacting antigen. Additional evidence that cross-reacting antibodies may have a pathogenic role is provided by studies of patients with the post-pericardiotomy syndrome in which a high incidence of heart-specific antibodies was noted (ENGLE et al. 1984). Further, 70% of these patients with heart-specific antibodies also had a fourfold or greater elevation of antibody titer to one of six viruses (four of which were coxsackie B subgroups). In contrast, of those patients with negative or intermediate titers of heart-specific antibody, only 5%–7% had elevated viral antibody titers. This type of evidence, particularly when considered together with Saegusa's identification of a Coxsackie B antigen that reacts uniquely with the human heart, makes a strong case for a pathogenic role for cross-reacting antibodies under certain circumstances. Thus, the challenge remains to explain how they are involved in the disease process, particularly at a molecular level.

It is worth remembering that cross-reactive antibodies are usually detected in the context of a broad immunological response that also involves cell-mediated immunity. In RF, for example, there is an increased cellular response to streptococcal antigens. Several subpopulations of lymphocytes can be identified both in heart valve inflammatory tissue and in the classical Aschoff lesion (KEMENY et al. 1987). The exact involvement of cell-mediated immunity and the role of these T-lymphocyte subpopulations in this disease are active areas of research in our laboratory.

Yet if cross-reactions between microbe and host are common and are involved in pathogenesis, why is related clinical disease so uncommon an occurrence? The resulting humoral and cellular disturbance leads us to the third necessary component, namely, genetic susceptibility.

In the battle of reciprocal adaptation between microbe and host, it has been suggested that the MHC has evolved as a "flight" from molecular mimicry of microbes, ensuring a population in which no two hosts look alike. This advantage, however, may be reduced by an increased susceptibility to autoimmune disease.

Could susceptibility to autoimmune disease in general (and RF in particular) be a function of one or a combination of HLA antigens? As stated above several studies have reported HLA associations with RF, albeit with low relative risks. This, however, does not necessarily exclude a genetic predisposition, since it may be that (a) the gene conferring disease susceptibility is only in linkage dysequilibrium with the putative HLA marker; and (b) the presently identified gene products of HLA alleles are themselves heterogeneous, and the disease may correlate with only one subtype. Evidence for a direct association between HLA antigens and RF is, nevertheless, lacking at present.

Of perhaps more relevance to RF has been the recent hypothesis of STROMINGER (1986), who suggests the involvement of two or more genes for autoimmune disease to occur. In the case of RF, the infection with group A streptococcus might generate a clone of MHC class II restricted, antigen-specific, cytotoxic T lymphocytes bearing a receptor idiotype that cross-reacts with a surface protein present on normal tissue. A second gene might determine a different polymorphism in the system at either a humoral or cellular level (i.e., an infrequent variant of the target protein). Although this theory is attractive, it must be noted that the heart reactive antibodies described in the sera of patients with RF react with cardiac tissue from both rheumatic and nonrheumatic individuals (ZABRISKIE et al. 1970).

An alternative hypothesis is that an RF susceptibility gene lies outside the MHC, since our studies (KHANNA et al. 1988) and those of PATARROYO et al. (1979) were unable to demonstrate an association with any HLA antigens so far defined. Preliminary studies of this marker on the B cells of RF families suggest that the marker is inherited in an autosomal recessive fashion, but this does not explain the relatively low concordance for disease in monozygotic twins, unless a second genetic polymorphism is involved. Streptococcal infection might provide the key to this second gene. Following infection, somatic mutation could result in the expansion of a B-cell population bearing this disease association marker, which is antigenically cross-reactive with a structure on the target organ. Support for this hypothesis comes from our observation that the D8/17 antigen is also present on skeletal muscle and cardiac tissue of non rheumatic individuals. This theory could explain the relatively low concordance for RF in monozygotic twins, since the second gene may depend on a somatic mutation following the inciting streptococcal infection. Future studies concerned with (a) the inheritance of this marker in multiplex rheumatic families, (b) the molecular basis of the gene product, and (c) the nature of the cross-reactive antigen will elucidate the exact nature of this marker in individuals with RF.

More important to our theme is the functional significance of this marker in the disease process. Does it control the *expression* of D8/17 positive cells? Is the antigen an internal image of the cross-reacting streptococcal antigen? Does the prolonged expression of the D8/17 antigen in RF patients (the increased number of D8/17-positive B cells is sometimes present 20–30 years after a single episode of RF) induce a "sensitized" state in these individuals permitting recurrence following streptococcal exposure later in life?

In the broader context, the study of biological mimicry has permitted us to gain important insights into a number of areas, yet many questions remain unanswered. For example, what is the molecular basis of many of these cross-reactions, and how does it result in disease? How do lymphocytes lose tolerance? What is the immunogenetic controlling element of these phenomena, and what is the molecular basis of susceptibility to infection? Microbe-host interactions provide an excellent laboratory in which to study these broad ranging issues in medical science.

References

Anastasiou-Nana MI, Anderson JL, Carlquist JF, Nanas JN (1986) HLA-DR typing and lymphocyte subset evaluation in rheumatic heart disease: a search for immune response factors. Am Heart J 112: 992–997

Ayoub EA, Barrett DJ, Maclaren NK, Krischer JP (1986) Association of class II human histocompatibility leucocyte antigens with rheumatic fever. J Clin Invest 77: 2019–2026

Barnett AH, Eff C, Leslie RDG, Pyke DA (1981) Diabetes in identical twins. Diabetologia 20: 87–93

Bendixin G (1968) Organ specific inhibition of the in vitro migration of neucocytes in human glomerulonephritis. Acta Med Scand 184: 99–103

Bessen D, Fischetti VA (1988) Influence of intranasal immunization with synthetic peptides corresponding to conserved epitopes of M protein on mucosal colonization by group A streptococci. Infect Immunol (in press)

Bronze MS, Beachey EH, Dale JB (1988) Protective and heart cross-reactive epitopes located within the amino-terminus of type 19 streptococci. J. Exp Med 167: 849–860

Cheadle WB (1889) Harvean lectures on the various manifestations of the rheumatic state as exemplified in childhood and early life. Lancet i: 821–832

Cunningham MW, Krisher K, Graves DC (1984) Murine monoclonal antibodies reactive with human heart and group A streptococcal membrane antigens. Infect Immun 46: 34–41

Cunningham MW, Hall NK, Krisher KK, et al. (1986) A study of anti-group A streptococcal monoclonal antibodies cross-reactive with myosin. J Immunol 136: 293–298

Cunningham MW, Swerlick RW (1986) Polyspecificity of anti-streptococcal murine monoclonal antibodies and their implications in autoimmunity. J Exp Med 164: 998–1012

Cunningham MW, McCormack JM, Talaber LR, Harley JB, et al. (1988) Human monoclonal antibodies reactive with antigens of the group A streptococcus and human heart. J. Immunol 141

Dale JB, Beachey EH (1982) Protective antigenic determinant and streptococcal M protein shared with sacrolemmal membrane protein of the human heart. J Exp Med 156: 1165–1176

Dale JB, Beachey EH (1985) Multiple cross-reactive epitopes of streptococcal M proteins. J Exp Med 161: 113–122

Dale JB, Beachey EH (1986a) Sequence of myosin-cross-reactive epitopes of streptococcal M protein. J Exp Med 162: 1785–1790

Dale JB, Beachey EH (1986b) Localization of protective epitopes of the amino terminus of Type 5 streptococcal M protein. J Exp Med 163: 1191–1202

Dale JB, Beachey EH (1987) Human cytotoxic T-lymphocytes evoked by group A streptococcal M proteins. J Exp Med 166: 1825–1835

Damian RT (1964) Molecular mimicry: antigen sharing by parasite and host and its consequences. Am Naturalist 98: 129–149

DiSciascio G, Taranta A (1980) Rheumatic fever in children. Am Heart J 99: 635–658

Dossetor JB, Schlaut J, Olson JA, Zabriskie JB (1978) Streptococcal infection as a cause of hyperacute renal allograft rejection. Trans Proceeds 10: 483–487

Dudding BA, Ayoub EM (1968) Persistance of streptococcal group A antibody in patients with rheumatic valvular disease. J Exp Med 128: 1081–1092

Ebringer A (1978) The link between genes and disease. New Scientist Sept: 86–867

Ellis RJ, Ebert PA, McCarty M, Zabriskie JB (1976) Prolongation of myocardial tissue allografts by pretreatment with streptococcal membrane. Transpl Proc VII: 355–359

Engle MA, Gay WA, Zabriskie JB, Senterfit LB (1984) Postpericardiotomy syndrome: 25 years' experience. J Cardiovasc Med April: 321–332

Faarber P, Capel PJ, Rigke PM, Vierwimden G, et al. (1984) Cross-reactivity of anti DNA antibodies with proteoglycans. Clin Exp Immunol 55: 402–412

Fillit HM, Read SE, Sherman RL, Zabriskie JB, et al. (1978) Cellular reactivity to alter glomerular basement membrane in glomerulonephritis. N Engl Med 298: 861–868

Fillit HM, Zabriskie JB (1982) Cellular immunity in glomerulonephritis. Am J Pathol 109: 227–243

Fillit HM, Damle SP, Gregory JD, Volin C, Poon-King T, Zabriskie JB (1984) Sera from patients with poststreptococcal glomerulonephritis contain antibodies to glomerula heparan sulphate proteoglycan. J Exp Med 161: 277

Fillit HM, McCarty M, Blake M (1986) Induction of antibodies to hyaluronic acid by immunization of rabbits with encapsulated streptococci. J Exp Med 164: 762–776

Finland M, Curnen E (1940) The agglutination of human erythrocytes in antipneumococcal sera. J Immunol 38: 457–478

Fischetti VA, Parry DAD, Trus BL, Hollingshead SK, et al. (1988) Conformational characteristics of the complete sequence of group A streptococcal M6 protein. Proteins: structure, function, and genetics 3: 60–69

Fitzsimmons EJ, Weber M, Lange CF (1987) The isolation of cross-reactive monoclonal antibodies: hybridomas to streptococcal antigens cross-reactive with mammalian basement membranes. Hybridoma 6: 61–69

Forsbery B, Johnson U (1987) Age at onset, sex distribution, and HLA antigen frequency in patients with primary glomerulonephritis progressing to terminal uraemia. An epidemiological survey. Scand J Urol Nephrol 21: 301–306

Freimer EH (1963) Studies of L forms and protoplasts of group A streptococci. J Exp Med 117: 377–399

Fujikawa S, OhKuni M, Lue HC (1987) Antibody titer to group A streptococcal polysaccharide in rheumatic fever and rheumatic heart disease. Jpn Circ J 51: 1347–1349

Goldstein I, Caravano R (1967) Determination of antigroup A streptococcal polysaccharide antibodies in human sera by an hemagglutination technique. Proc Soc Exp Biol Med 124 (4): 1209 to 1212

Goldstein I, Rebeyrotte P, Parlebas J, Halpern B (1968) Isolation from heart valves of glycopeptides which share immunological properties with *Streptococcus haemolyticus* group A polysaccharides. Nature 219: 866–868

Goroncy-Bearns P, Dale JB, Beachy EH, Opferkuch W (1987) Monoclonal antibody to human renal glomeruli cross-reacts with streptococcal M protein. Infect Immun 55: 2416–2419

Gowrishankar R, Agarwal SC (1980) Leucocyte migration inhibition with human heart valve glycoproteins and group A streptococcal ribonucleic proteins in rheumatic heart diesease and post streptococcal glomerulonephritis. Clin Exp Immunol 39: 519–528

Hafez M, Chakravarti A, El-Shennawy S, El-Morsi Z, El-Sallab SH, Al-Yonbary Y (1985) HLA antigens and acute rheumatic fever. Genetic Epidem 2: 273–282

Hirata AA, Terasaki PI (1970) Cross-reactions between streptococcal M protein and human transplantation antigens. Science 168: 1095–1098

Hirsch JC, Church AB (1960) Studies of phagocytosis of group A streptococci by polymorphonuclear leucocytes in vitro. J Exp Med 111: 309–322

Holm SE (1967) Precipitogens in B-hemolytic streptococci and some related human kidney antigens. Acta Pathol Mcrobiol Scand 70: 79–94

Holm SE, Christensen P, Shalen C (1978) Alternate interpretation of cross-reactions between streptococci and human tissue. In: Pathogenic streptococci. Reedbooks, Surrey

Hosein B, McCarty M, Fischetti VA (1979) Amino acid sequence and physiological similarities between streptococcal M protein and mammalian tropomyosin. Proc Natl Acad Sci USA 76: 3765–3768

Humphreys JH (1943) Antigenic properties of hyaluronic acid. Biochem J 37: 460–463

Husby G, van de Rijn I, Zabriskie JB, Abdin ZH, Williams RC Jr (1976) Antibodies reacting with cytoplasm of subthalmic and caudate nuclei neurons in chorea and acute rheumatic fever. J Exp Med 144: 1094–1110

Jones DB (1951) Inflammation and repair of the glomerulus. Am J Pathol 27: 991–1009

Kaplan MH (1963) Immunologic relation of streptococcal and tissue antigens. J Immunol 90: 595–606

Kaplan MH, Frengley JD (1969) Autoimmunity to the heart in cardiac disease. Current concepts of the relationship of autoimmunity to rheumatic fever, post cardiotomy, post infarction syndromes, and cardiomyopathies. Am J Cardiol 24: 459–473

Kemeny E, Grieve T, Marcus R, Sarelli P, Zabriskie JB (1987) Identification of mononuclear cells and T cell subsets in rheumatic valvulitis. Xth Lancefield international symposium on streptococci and streptococcal diseases

Kendall F, Heidelberger M, Dawson M (1937) A serologically inactive polysaccharide elaborated by mucoid strains of group A hemolytic streptococcus. J Biol Chem 118: 61–82

Khanna AK, Williams RC Jr, Gibofsky A, Buskirk D, Crow M, Menon A, Fotino M, Reid H, Poon-King T, Rubinstein P, Zabriskie J (1989) The presence of a non-HLA B-cell antigen in rheumatic fever patients and their families as defined by monoclonal antibody. J Clin Invest 83: 1710–1736

Kingston D, Glynn LE (1971) A cross-reaction between *Streptococcus pyogenes* and human fibroblasts, endothelial cells and astrocytes. Immunology 21: 1003–1016

Kraus W, Beachey EH (1988) Renal autoimmune epitope of group A streptococci specific by M protein tetrapeptide Ile-Arg-Leu-Arg. Proc Natl Acad Sci USA 85: 4516–4520

Krisher K, Cunningham MW (1985) Myosin: a link between streptococci and heart. Science 277: 413–415

Lancefield RC (1962) Current knowledge of the type specific M antigens of group A streptococci. J Immunol 89: 307–313

Lange K, Gold MMA, Weiner D, Simon V (1948) Autoantibodies in human glomerulonephritis. J Clin Invest 28: 50–55

Layrisse Z, Rodriguez-Iturbe B, Garcia-Ramirez R, Rodriguez A, Tiwari J (1983) Family studies of the HLA system in acute post-streptococcal glomerulonephritis. Hum Immunol 7: 177–185

Lui CT, McCrory WW (1958) Autoantibodies in human glomerulonephritis and nephrotic syndrome. J Immunol 81: 492–498

Manjula BM, Fischetti VA (1980) Tropomyosin-like serum residue periodicity in three immunologically distinct streptococcal M protein and its implication for the antiphagocytic property of the molecule. J Exp Med 151: 695–708

Markowitz AS, Armstrong JS, Kushner DS (1960) Immunological relationships between the rat glomerulus and nephritogenic streptococci. Nature 187: 1095–1097

Markowitz AS, Lange CF (1962) Streptococcus related glomerulonephritis. J Lab Clin Med 60: 1001–1002

Markowitz AS, Lange CF (1964) Streptococcal related glomerulonephritis. J Immunol 92: 565–575

Markowitz AS (1969) Streptococcal-related glomerulonephritis in the rhesus monkey. Trans Proc 1: 985–991

McCarty M (1970) The streptococcal cell wall. The Harvey lectures series 65: 73–96

Meyer K (1958) Chemical structure of hyaluronic acid. Fed Proc 17: 1075–1077

Meyer K, Palmer JW (1936) On glycoproteins: the polysaccharides of vitreous humor and of umbilical cord. J Biol Chem 107: 689–699

Neu N, Beisel KW, Traystman MD, Rose NR, et al. (1987a) Autoantibodies specific for the cardiac myosin isoform are found in mice susceptible to coxsackievirus B_3-induced myocarditis. J Immunol 138: 2488–2492

Neu N, Craig SW, Rose NR, et al. (1987b) Coxsackievirus induced myocarditis in mice: cardiac myosin autoantibodies do not cross-react with the virus. Clin Exp Immunol 69: 566–574

Pancholi VK, Fischetti VF (1987) Isolation and characterization of cell associated region of group A streptococcal M protein. J Bacteriol 170: 2618–2624

Patarroyo ME, Winchester RJ, Vejerano A, Gibofsky A, et al. (1979) Association of a B cell alloantigen with susceptibility to rheumatic fever. Nature 278: 173–174

Perlman P, Hammarstrom S, Lagercrantz R, Gustafsson BE (1965) Antigen from colon of germfree rats and antibodies in human ulcerative colitis. Ann NY Acad Sci 124: 377–394

Phillips GM, Flicker PF, Cohen C, Manjula BM, Fischetti VA (1981) Streptococcal M protein: alpha-helical coiled-coil structure and arrangement on cell surface. Proc Nat Acad Sci USA 78: 4689–4693

Prioli RP, Ordovas JM, Rosenberg I, Schaefer EJ, Pereira MEA (1987) Similarity of cruzin, an inhibitor of *Trypanosoma cruzi* neuraminidase, to high-density lipoprotein. Science 238: 1417–1419

Quinn RW, Singh KP (1957) Antigenicity of hyaluronic acid. Biochem J 95: 290–301

Rapaport FT, Chase RM Jr (1964) Homograft sensitivity induction by group A streptococci. Science 145: 407–408

Jhingan B, Reddy KS, Taneja V, Vaidya MC, Bhatia ML (1986) HLA, blood groups and secretor status in patients with established rheumatic fever and rheumatic heart disease. Tissue Antig 27: 172–178

Rapaport RT, Markowitz AS, McClusky RT, Hanaoka T, Shimada T (1969) Induction of renal disease with antisera to group A streptococcal membrane. Trans Proc 1: 981–984

Read SE, Zabriskie JB, Fischetti VA, Utermohlen V, et al. (1974) Cellular reacting studies to streptococcal antigens I. Migration inhibition studies in patients with streptococcal infections and rheumatic fever. J Clin Invest 54: 439–450

Reid HFM, Read SE, Poon-King T, Zabriskie JB (1980) Lymphocyte response to streptococcal antigens in rheumatic fever patients in Trindad. In: Read SE and Zabrieskie JB (eds) Streptococcal diseases and the immune response. Academic, New York, pp 681–693

Rocklin RE, Lewis EJ, David JB (1970) In vitro evidence for cellular hypersensitivity to glomerular basement membrane antigens in human glomerulonephritis. HEJM 283: 497–501

Rose NR, Wolfgram LJ, Herskowitz A, Beisel KW (1986) Postinfectious autoimmunity: two distinct phases of coxsackie B_3-induced myocarditis. Ann NY Acad Sci 475: 146–156

Rose NR, Beisel KW, Herskowitz A, Neu N, et al. (1987) Cardiac myosin and autoimmune myocarditis. In: Autoimmunity and autoimmune disease. Wiley, Chicester

Roses DF, Zabriskie JB, Rapaport FT (1973) White graft rejection after streptococcal and staphylococcal infection in man. Trans Proc V: 487–489

Rothbard S (1948) Protective effect of hyaluronidase and type-specific anti-M serum on experimental group A streptococcus infections in mice. J Exp Med 88: 325–342

Rowley D, Jenkins CR (1962) Antigenic cross-reaction between host and parasite as a possible cause of pathogenicity. Nature 193: 151–154

Saegusa J, Prabhakar BS, Essani K, McClintock PR, et al. (1986) Monoclonal antibody to coxsackie virus B_4 reacts with myocardium. J Infect Dis 153: 372–373

Sandson J, Hammerman D (1962) Isolation of hyaluronate protein from human synovial fluid. J Clin Invest 41: 1817–1830

Sasazuki T, Iwamoto I, Tsuchida H (1979) HLA and acute poststreptococcal glomerulonephritis. N Engl J Med 301 (21): 1184–1185 (letter)

Sargent SJ, Beachey EH, Corbett CE, Dale JB (1987) Sequence of protective epitopes of streptococcal M proteins shared with cardiac sarcolemmal membranes. J Immunol 139: 1285–1290

Schwimmbeck PL, Yu DTY, Oldstone MBA (1987) Autoantibodies to HLA B27 in the seren of patients with ankylosing spondylitis and Reiter's syndrome. J Exp Med 166: 173–181

Seager K, Bashir HV, Geczy AF, Edmonds J, De Vere-Tyndall A (1979) Evidence for a specific B27 associated cell surface marker on lymphocytes of patients with ankylosing spondylitis. Nature 277: 68–70

Seastone CV (1939) The virulence of group C hemolytic streptococci of animal origin. J Exp Med 70: 361–378

Springer GF, Williamson P, Brandes WC (1961) Blood group activity of gram-negative bacteria. J Exp Med 113: 1077–1093

Stollerman GH, Rytel M, Ortiz J (1963) Accessory plasma factors involved in the bactericidal test for type-specific antibody to group A streptococci. J Exp Med 117: 1–17

Strominger JL (1986) Biology of the human histocompatibility leukocyte antigen (HLA) system and a hypothesis regarding the generation of autoimmune diseases. J Clin Invest 77: 1411–1415

Taranta A, Torosdag S, Metrakos JD, Jegier W, et al. (1959) Rheumatic fever in monozygotic and dizygotic twins. Circulation 20: 778–792

Tauber JW, Falk JA, Falk FE, Zabriskie JB (1976a) Nonspecific complement activation by streptococcal structures I. Re-evaluation of HLA cytotoxicity inhibition. J Exp Med 143: 1341–1351

Tauber JW, Polley MJ, Zabriskie JB (1976b) Nonspecific complement activation by streptococcal structures. II. Properdin-independent initiation of the alternate pathway. J Exp Med 143: 1352–1366

Tauber JW, Zabriskie JB (1980) Streptococcal structures and histocompatibility markers: did HLA evolve as an escape from microbial mimicry? In: Read SE and Zabriskie JB (eds) Streptococcal diseases and the immune response. Academic, New York, pp 607–618

Underhill CB (1982) Naturally occurring antibodies which bind hyaluronate. Biochem Biophys Res Comm 108: 1488–1494

van de Rijn I, Triscott MX (1984) Analysis of group A and C streptococcal membrane proteins. In: IX Lancefield international symposium on recent advances in streptococci and streptococcal diseases. Reedbooks, Bracknell

van de Rijn I, Bjleiweis AS, Zabriskie JB (1976) Antigens in *Streptococcus mutans* cross-reactive with human heart muscle. J Dental Res 55: C59–C64

van de Rijn I, Zabriskie JB, McCarty M (1977) Group A streptococcal antigens cross-reactive with myocardium: purification of heart-reactive antibody and isolation and characterization of the streptococcal antigen. J Exp Med 146: 579–599

van Eden W, Holoshitz J, Nevo Z, Frenkel A, Klajman A, Cohen R (1985) Arthritis induced by a T-lymphocyte clone that responds to mycobacterium tuberculosis and to cartilage proteoglycans. Proc Natl Acad Sci USA 82: 5117

Williams A, Eldrige R, McFarland H, Houff S, Krebs H, McFarlin DE (1980) Multiple sclerosis in twins. Neurology 30: 1139–1147

Williams RC, van de Rijn I, Reid HMF, Poon-King T, et al. (1981) Lymphocytic cell subpopulations during acute post streptococcal glomerulonephritis: cell surface antigens and binding of streptococcal membrane antigens and cross-reactions protein. Clin Exp Immunol 46: 397–405

Wilson MG, Schweitzr MD, Lubschez R (1943) The familial epidermiology of rheumatic fever. J Pediatr 22: 468–482

Zabriskie JB (1971) The role of streptococci in human glomerulonephritis. J Exp Med 134: 180–192

Zabriskie JB (1985) Rheumatic fever: the interplay between host genetics and microbe. Circulation 71: 1077–1086

Zabriskie JB, Freimer EH (1966) An immunological relationship between the group A streptococcus and mammalian muscle. J Exp Med 124: 661–678

Zabriskie JB, Hsu KC, Seegal BC (1970) Heart reactive antibody associated with rheumatic fever: characterization and diagnostic significance. Clin Exp Immunol 7: 149–159

A Cartilage-Mimicking T-Cell Epitope on a 65K Mycobacterial Heat-Shock Protein: Adjuvant Arthritis as a Model for Human Rheumatoid Arthritis

W. van Eden, E. J. M. Hogervorst[1], E. J. Hensen[1], R. van der Zee[2], J. D. A. van Embden[2], and I. R. Cohen[3]

1 Introduction   27
2 Mycobacteria and Selective Recognition by the Host: From Peaceful Coexistence to Immunopathology   28
3 Mycobacterial Antigens May Induce Arthritis in Rats: The Experimental Model of Adjuvant Arthritis   30
4 Antigenic Mimicry Between Mycobacteria and Cartilage: A Conserved 65 K Heat-Shock Protein   32
5 Mycobacterial Mimicry and Rheumatoid Arthritis   35
6 Mimicry and the Restoration of Self-Tolerance   38
References   40

1 Introduction

The way the immune system evolved has been commonly thought to have been influenced during evolution through selective pressure exerted by microbial invaders. So, immune recognition and the subsequent immune response have become significant means of the host to combat the attack of exogenous invaders. Since, however, the exogenous microbial world presents itself with a wealth of antigenic variety on each single organism, immune recognition can be selective at the antigen level. Teleologically, an unselected response would be a very inefficient maneuver inevitably leading to jamming of the system. Moreover, one could envisage more pertinent reasons for being selective, among them that responding to certain antigens might jeopardize the maintenance of self integrity. This means that a response directed to such antigens might cause uncontrollable disturbances in the balance of elements that interact in the immunological network and might lead to the ultimate development of immunopathology. This could happen when the exogenous stimulator antigen bears a resemblance to self molecules, a situation called antigenic mimicry.

It is clear that reactivity to self molecules is normally avoided or carefully controlled. In other words, tolerance for self seems to be the rule thanks either to holes in

---

[1] Dept. of Infectious Diseases and Immunology, Veterinary Faculty, Utrecht University, Utrecht, The Netherlands
[2] Dept. of Bacteriology, National Institute of Public Health and Environmental Hygiene, Bilthoven, The Netherlands
[3] Dept. of Cell Biology, The Weizmann Institute of Science, Rehovot, Israel.

the recognition repertoire, to active immune regulation, or to a combination of both at the same time as demonstrated in T-cell nonresponsiveness for some exogenous antigens (SERCARZ 1987). Furthermore, again comparable with the immune response to exogenous antigen, tolerance to self antigens follows the rules of MHC-restricted T-cell recognition (MATZINGER et al. 1984; RAMMENSEE and BEVAN 1984). Altogether, it appears that T cells play an essential role in conserving self tolerance. The MHC molecules as antigen-presenting structures have a crucial task in the selection of epitopes to which the immune system responds or does not respond, thereby maintaining a safe balance between tolerance and responsiveness in the network. This task apparently has shaped the MHC into an extremely polymorphic genetic system during evolution under the combined pressure of both the outside world and the self. It seems that as a consequence of this pressure from time to time the immune response has to compromise and is forced, seemingly, to choose an "incorrect" epitope to respond to, resulting in disease. Microbes, successfully exploiting their mimicry with host self antigens, might hamper the immune response in an infectious disease. Alternatively, this same mimicry might well be at the origin of a specific loss of self-tolerance. Although in the latter case the problem of infection can be effectively avoided, the host might pay the bill by developing immune reactivity against a self-mimicking epitope that has only an incomplete fit with some receptors interacting in the network. The ensuing imbalance of the response might easily lead to some sort of overcompensation in the form of reactivity directed against self.

Such aberrations of immune regulation are exemplified by the host in its confrontation with omnipresent organisms such as mycobacteria in leprosy and also in the experimental disease adjuvant arthritis (AA) (VAN EDEN et al. 1986). This latter autoimmune disease model can be elicited by immunization with a genuine exogenous antigen. As will be discussed later, the disease has been found to originate from true molecular mimicry, and it is just through immunization with the very same mimicking inducer antigen that, under some circumstances, protection against this autoimmune disease can be achieved. Therefore, the model of AA might teach us strategies to follow when developing new therapeutic means to combat autoimmune disease, at least when disease is the consequence of a response confused in its discrimination between self and nonself due to molecular mimicry.

2 Mycobacteria and Selective Recognition by the Host: From Peaceful Coexistence to Immunopathology

Diseases caused by mycobacteria are characterized by a spectrum of clinical appearances determined by the quality of the individual host immune response. Leprosy serves in this respect as a well-studied "model disease." The etiologic agent, *Mycobacterium leprae*, is like other mycobacteria a remarkably innocuous microorganism in itself. The host is capable of carrying large quantities of this organism intracellularly in its tissues without too much trouble. This is seen in what is called lepromatous leprosy, the form of the disease that can develop in the absence of or under down-regulation of specific cell-mediated immunity. When, however, specific T cell immunity is generated, the disease sometimes develops into tuberculoid leprosy, in which most of the organisms are eliminated (RIDLEY and JOPLING 1966). The price the host has to pay for this effective elimination of the invader is, in this case, immunopathology. The inflammatory response forms granulomas characteristically located in the close vicinity of peripheral nerves with irreversible nerve damage as a consequence. Although the majority of chronically diseased individuals develop some variable degree of reactivity to the causative agent and show therefore a form of disease intermediate between lepromatous and tuberculoid leprosy, a significant proportion of individuals apparently make an irreversible choice between either responding or not responding to the organism. In both cases, the organism is the same; no antigenic variants have ever been observed.

The obvious question is how this choice is made. The analysis of a large group of multi-case families has shown that the MHC is part of the answer. The segregation patterns of parental HLA-haplotypes in these families occurs in a nonrandom manner, so that patients suffering from the same form of leprosy in a family tend to have inherited identical HLA-haplotypes (Fig. 1) (DE VRIES et al. 1976; VAN EDEN and DE VRIES 1984; VAN EDEN et al. 1985a). The most likely interpretation of this is that the MHC type of the host plays a decisive role in the selection of mycobacterial antigens to be presented to regulatory T cells. In vitro experimental data support this interpretation.

In lepromatous leprosy the cellular anergy is clearly caused by specific T-cell suppression (MEHRA et al. 1982), and some specific mycobacterial antigens have been found responsible for the induction of the suppressive response (MEHRA et al. 1984). Furthermore, presentation of mycobacterial antigens via the HLA-DQ1 molecule has been shown to have a suppressive influence on HLA-DR-restricted anti-mycobacterial T-cell responses (VAN EDEN et al. 1984a). This regulatory role of HLA-DQ1 in anti-mycobacterial T-cell responses in vitro is the possible explanation for the increased frequency of HLA-DQ1 among lepromatous leprosy patients in some populations (OTTENHOFF et al. 1984). Similarly, the failure of HLA-DR3 to present certain *M. leprae*-specific antigens to immune T cells (VAN EDEN et al. 1984b) might be the *in vitro* correlate for the observed associations between HLA-DR3 and both the delayed type hypersensitivity (DTH) reactivity to so-called mycobacterial common antigens in skin test tuberculins (VAN EDEN et al. 1983) and the HLA-DR3-encoded predisposition to develop the tuberculoid form of leprosy (VAN EDEN et al. 1982). In the same ethnic group whose HLA-DR3 was found to be associated with

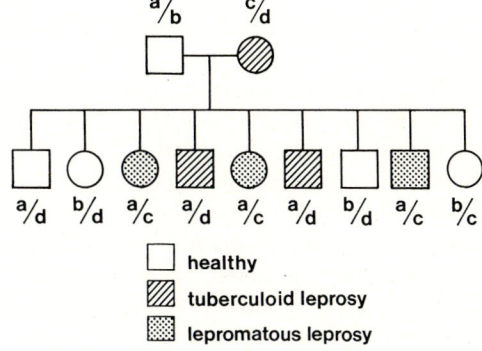

Fig. 1. Segregation of HLA-haplotypes in a Venezuelan multiple case family with leprosy. Children with the same form of leprosy inherited identical haplotypes, indicating a role for HLA in determining leprosy type. (Family was taken from VAN EDEN et al. 1985a)

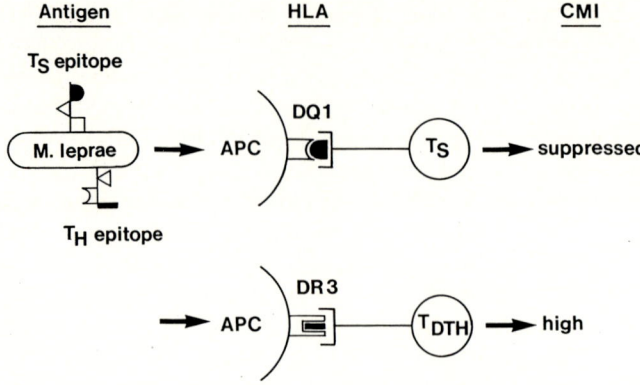

Fig. 2. Schematic representation of (HLA) molecules present on antigen-presenting cells (*APC*) selecting mycobacterial antigens. Depending on the epitope presented, functionally distinct T-cell subsets are activated, determining the clinical outcome of cell-mediated immunity (*CMI*) in leprosy

tuberculoid leprosy, this HLA specificity was virtually absent among patients with lepromatous leprosy. Thus, HLA-DR3 prevented infected individuals from developing T-cell anergy or down-regulation with respect to the mycobacterial invader. All HLA-DR3-positive individuals made a choice for T-cell reactivity with immuno-pathological sequelae. In most ethnic groups, HLA-DR3 has been associated with various autoimmune diseases. Although it might be a mere coincidence, also in the case of leprosy, HLA-DR3 was associated with the immunopathological type of response. Whether true autoimmunity is involved in the nerve damage in this form of leprosy (CRAWFORD et al. 1974) is still debatable. Nevertheless, leprosy illustrates the point that, depending on the genetic make-up of the host, the immune system upon its encounter with an exogenous invader selects a particular epitope from the large number of possible antigens. It is this selection that predetermines the type of response that follows, either a suppressive one allowing a more or less peaceful coexistence of host and parasite or an agressive one that can lead to immune damage to host self tissues (Fig. 2).

3 Mycobacterial Antigens May Induce Arthritis in Rats: The Experimental Model of Adjuvant Arthritis

Mycobacterial antigens are capable of inducing not only immunopathology such as granuloma formation, as seen in leprosy and tuberculosis, but also arthritis. This is seen in rats after intracutaneous immunization with heat-killed mycobacteria suspended in mineral oil, a preparation known as complete Freund's adjuvant (PEARSON 1956). Therefore, the experimental disease is called adjuvant arthritis (AA). Genetic predisposition exists and is caused, at least in part, by the host MHC (BATTISTO et al. 1982). This experimentally induced arthritis has been studied, because of its striking histopathological similarity, as a model for human rheumatoid arthritis. The arthritis is especially prominent in the small joints of the extremities and is characterized by

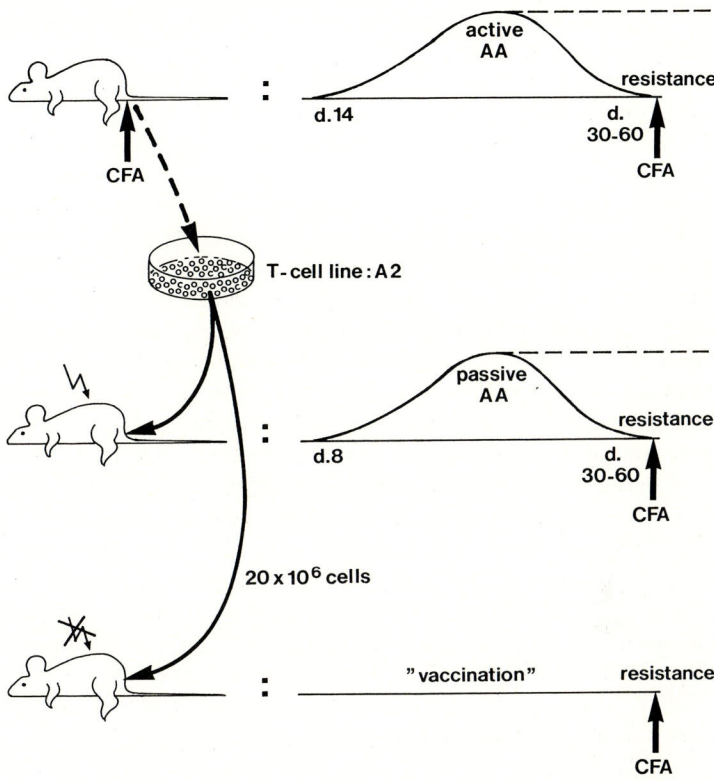

Fig. 3. Schematic representation of adjuvant arthritis (*AA*) induced by either immunization with complete Freund's adjuvant (*CFA*) or T-cell line A2 in irradiated or nonirradiated recipient rats. Both CFA- and A2-induced disease show spontaneous remission at 30–60 days after immunization. A CFA challenge given after remission does not reactivate arthritis. Apparently resistance to developing AA has been obtained. Similar resistance has been seen to arise in the absence of AA after inoculation with A2 in nonirradiated recipients. Thus A2 may vaccinate against AA

inflammation of synovium, pannus formation, destruction of cartilage, and bone erosion that may ultimately lead to bony ankylosis and deformation of the affected joints. The question whether or not the "laboratory curiosity" of AA is a faithful animal copy of any human rheumatological disease remains difficult to answer at this stage. Without any doubt, however, the entity of AA is suited for the study of how immunization to bacterial antigens can lead to an autoimmune form of arthritis. Clearly this is relevant for rheumatology in particular and autoimmunity in general.

The autoimmune nature of the disease was suspected initially, because the disease was transferable to naive recipient rats by lymphocytes obtained from affected rats (PEARSON and WOOD 1964). However, elegant proof of the autoimmune nature of the disease was given by Holoshitz and coworkers (HOLOSHITZ et al. 1983), when they succeeded in isolating from such immunized Lewis rats the so-called A2 T-cell line, which transferred the disease to naive, irradiated, syngeneic, recipient rats (Fig. 3). This T-cell line had been selected and maintained by repeated *in vitro* stimulation in the presence of irradiated antigen-presenting cells and whole mycobacteria as an antigen.

Susceptibility to developing AA after mycobacterial immunization suggested that the arthritis-inducing antigen in mycobacteria is easily selected by T cells in the context of Lewis MHC. So, although the inducing antigen was unknown, it was expected that *in vitro* the T cells would be able to select the antigen among the variety of antigens present in whole mycobacteria. This reasoning turned out to be correct, when one of the developing lines was found to be arthritogenic in irradiated syngeneic recipients. The protocol of selection was based on earlier successes of Ben-Nun (BEN-NUN and COHEN 1982) in inducing experimental allergic encephalomyelitis (EAE) with a helper T-cell line, Z1a, reactive to the basic protein of myelin. Similar to the EAE-causing T-cell line Z1a, the arthritogenic A2 line was found to have the phenotype of helper or DTH T cells. In contrast to EAE, however, the A2 line induced disease only in recipients that had been sublethally irradiated. Apparently, host resistance mechanisms had to be weakened before inoculation of the line cells could lead to a pernicious shift in the balance. Yet, in the nonirradiated recipients inoculation of A2 was found to induce resistance to active, mycobacteria-induced AA. Subsequent subcloning by limiting dilution of the A2 cell line revealed the presence within A2 of functionally contrasting T-cell populations, explaining both the disease-inducing and the protective qualities of A2 depending on the circumstances. One subclone, however, was found to be exclusively arthritogenic and not protective. This clone was called A2b (HOLOSHITZ et al. 1984). Having isolated a clone with a single receptor specificity, which was virulently arthritogenic, we had a perfect tool to search for antigens critical to the origin of AA.

4 Antigenic Mimicry Between Mycobacteria and Cartilage: A Conserved 65K Heat-Shock Protein

After intravenous inoculation into syngeneic recipient rats, A2b was able to cause arthritis within a period of 5 days in irradiated recipients. The most likely interpretation of this would be that A2b interacted directly with an antigen located in joint tissues. This reasoning seemed to be correct. We knew already that A2b, in contrast to the original line A2, did not respond to collagen type II (HOLOSHITZ et al. 1984). However, when we cultured A2b in the presence of crude cartilage extracts enriched for proteoglycans, significant proliferative responses were seen (VAN EDEN et al. 1985b). Similar responses were subsequently obtained with culture supernatants of in vitro grown chick chondrocytes, which were known to produce proteoglycans, and also with synovial fluids obtained from human joints. The exact definition of the antigen within proteoglycans was technically hampered by the lack of chemically pure and undenatured subfractions of proteoglycans. Nevertheless, from the fact that heparan sulfate, keratan sulfate, and hyaluronic acid as such were negative and that relatively good responses were seen with a core protein-enriched subfraction, the antigen was likely to be associated with the core protein, the structure that carries the side chains, or the link protein, which is the protein that connects the core protein to the hyaluronic acid backbone of the proteoglycan molecule (CAPLAN 1984).

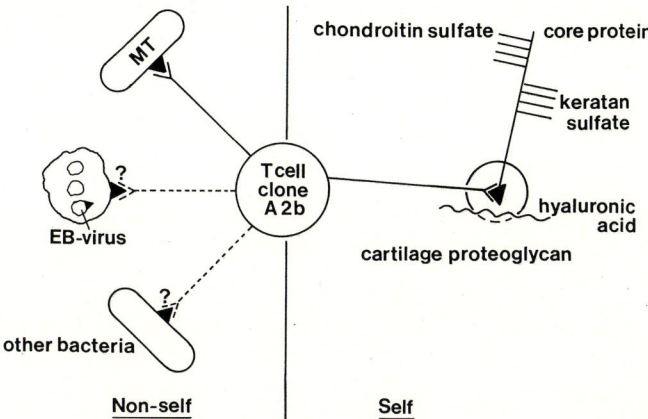

Fig. 4. Arthritogenic T cell clone A2b responds to both *M. tuberculosis* (*MT*) and cartilage components because of a mimicry between *M. tuberculosis* and a self component present most likely in the core protein or link protein of cartilage proteoglycan molecules. The mimicry epitope might be present on additional exogenous molecules expressed by organisms such as Epstein-Barr virus (*EB*) or other bacteria

Additional evidence for the cross-reactivity of A2b with proteoglycan was obtained from in vivo DTH reactivity not only to *M. tuberculosis* but also, although to a lesser degree, to our proteoglycan preparation (VAN EDEN et al. 1985b). Similar DTH responses to proteoglycans were obtained in *M. tuberculosis*-immunized rats suffering from AA.

These findings clearly indicate structural mimicry of a T-cell epitope in mycobacteria and cartilage proteoglycans (Fig. 4). Mimicry at the level of T cells between bacterial antigens and cartilage has also been shown by others (VAN DEN BROEK et al. 1988). Since, however, the mimicry of AA was defined by the arthritogenic clone A2b itself, it was likely that this mimicry indeed could incite the immune attack on joints seen in AA (VAN EDEN et al. 1987).

In order to identify the mimicking mycobacterial antigen, we started to screen clone A2b for reactivity against mycobacterial proteins obtained from an *E. coli* expression library, made by J. E. R. Thole of the National Institute of Public Health and Environmental Hygiene in Bilthoven (THOLE et al. 1985). A2b was found to recognize a 65K protein originating from *M. bovis* BCG (VAN EDEN et al. 1988). To identify the epitope more exactly, fragments of the 65K protein were obtained from deletion mutants of the gene for the protein, or from deletion mutants of the gene after fusion with the beta-galactosidase gene. From these fragments, it was discovered that A2b recognizes an epitope in the region between amino acid residues 171 and 234 of the 540-amino acid protein, an area small enough to span with synthetic peptides. The smallest synthetic peptide that stimulated A2b was a nonapeptide on positions 180–188. The amino acid sequence of this nonapeptide is shown in Table 1. Recently, by an alternative way of synthesis, all possible nonapeptides between positions 170 and 205 were made, in addition to peptides with deletions and substitutions at all positions. From this, it turned out that all the seven N-terminal amino acids were essential for stimulation of A2b, narrowing our epitope to 180–186 (VAN DER ZEE

Table 1. Sequence homologies of the 180–188 mycobacterial nonapeptide

180 188	
T F G L Q L E L T	65K of mycobacteria
T – – – – L E L –	Link protein of proteoglycan
T F G L Q – – – T	Epstein-Barr virus (BPLF-1)
– – – – Q L E L –	HLA-DQ 3 (beta-1 domain)
– – – L Q L E L –	Lamine
T F G L – L E – –	Hepatitis A virus

et al., 1989). Herewith, the mycobacterial mimicry epitope had been elucidated at the amino acid level. Since the cartilage counterpart epitope had been tentatively located in the protein part of the proteoglycan molecule, known amino acid sequences of core proteins and link proteins were compared with the nonapeptide. The best resemblance found was with a published sequence present in the rat link protein (NEAME et al. 1986); four out of the nine amino acids were identical. Whether this resemblance reflects the mimicry we were looking for remains to be seen. Alternatively, the homology could well be with a part of the protein that has not been sequenced as yet, or the mimicry could be based on structural similarity and not sequence identity (COHEN 1988).

Interestingly, our analysis of amino acid homologies showed identity for six of the nine amino acids with a hypothetical Epstein-Barr (EB) virus protein, BPLF-1. The synthetically made peptide based on this EB virus sequence did stimulate the A2b clone. This offered the possibility that potentially the same mimicry might originate from an epitope expressed by cells infected with EB virus. This, of course, might be of particular interest because of the variety of independant findings that allude to a possible role of EB virus in the pathogenesis of rheumatoid arthritis (e.g., Fox et al. 1986; ROUDIER et al. 1988). Another identity was found with a sequence that has been described as an HLA-DQ-specific sequence (Table 1). This concerns the four amino acids QLEL found in HLA-DQ3, a marker more strongly associated with rheumatoid arthritis in humans than HLA-DR4 (see later). Although we do not know whether the Lewis MHC carries this sequence, it might be a case of sequence identity with the "presenting" MHC molecule, as was suggested by GUILLET et al. (1987).

The nature of the mycobacterial antigen that carries the mimicking epitope is a special one. It is a member of a group of well-conserved protein molecules, now known as heat-shock proteins (YOUNG et al. 1987; SHINNICK et al. 1988). Homologues of the mycobacterial 65K protein are present among all mycobacteria and also other gram-positive and gram-negative bacteria (THOLE et al. 1988), including bacterial species frequently associated with arthritic conditions such as *Neisseria*, *Streptococcus*, *Shigella*, *Yersinia*, *Klebsiella*, *Campylobacter*, and others. The 65K protein must be one of their common components. It is, therefore, tempting to speculate that their possible significance for arthritic disease is through sharing the cartilage-mimicking epitope(s) on the protein. The presence of the 65K heat-shock protein also extends to eukaryotes and even mammalian cells (MCMULLIN and HOLBERG 1988). The hamster 65K homologue obtained from CHO cells has been recently cloned in *E. coli*. Its sequence showed about 60% identity with the mycobacterial 65K molecule (R. Young, personal communication). This is the highest homology ever found between a bacterial and a mammalian protein.

That animal and human cells were harboring a 58–65K molecule resembling its mycobacterial counterpart had been observed previously when we detected, in Western blots, a 65K-like molecule with a polyclonal antiserum and monoclonal antibodies raised against the mycobacterial 65K molecule. Interestingly enough, these antibodies also identified a 65K-like molecule in an extract of rat cartilage (Van der Zee and Van Embden, unpublished observation). Whether, in the latter case, the molecule detected was the human homologue of the 65K heat-shock protein is not yet known. Be that as it may, the existence of such exceptionally well-conserved proteins might easily be hazardous to the immune system.

Short stretches of identical sequence in molecules shared by bacteria and the host may offer many possibilities for dangerous mimicry. An immune response directed at these molecules is very hard to shield from cross-responding to self. Nevertheless, for instance, as in the case of the 65K molecule on mycobacteria, the molecule is regarded as behaving like an immunodominant antigen. In BCG-vaccinated mice and humans or tuberculosis patients, both antibodies and T cells with specificity for the 65K molecule are abundant (EMMRICH et al. 1986; KAUFMANN et al. 1987; THOLE et al. 1987). Also, from leprosy patients T cells with specificity for the molecule have been obtained (VAN SCHOOTEN et al. 1989). This is difficult to explain unless one assumes that the immune response is mainly concerned with those parts of the molecule that are nonhomologous with the 65K self molecule in the host.

Remarkably enough, in the Lewis rat the T-cell response against the 65K molecule seems to concentrate itself on the 180–188 mimicry epitope. Recently, A. Noordzij of our group generated a T-cell line after *in vivo* immunization and *in vitro* restimulation with the mycobacterial 65K antigen. Upon testing, this new line too turned out to be specific for the 180–188 sequence. So, it seems that the Lewis MHC tends to select this single epitope when confronted with either the 65K molecule or even whole *M. tuberculosis*. It is tempting, therefore to speculate that this phenomenon accounts for the remarkable susceptibility of Lewis rats to AA and that AA is exclusively dependent on the presence of this single epitope in mycobacteria that mimics a cartilage-associated molecule.

5 Mycobacterial Mimicry and Rheumatoid Arthritis

Although in Western countries the incidence of mycobacterial diseases such as leprosy and tuberculosis has declined over the past decades, nonetheless mycobacteria are still widely present, and daily contact with them occurs to all of us. The obvious question one might ask is, given the mimicry of mycobacteria with cartilage, whether arthritis in humans may be a consequence of the same mimicry. Although there certainly is no definite answer to this at present, some findings are suggestive (Table 2). A situation relatively comparable to AA in the rat might occur when individuals are treated with BCG for cancer immunotherapy. In some patients, arthritis is seen to develop (TORISU et al. 1978).

To investigate whether patients suffering from rheumatoid arthritis had some augmented T-cell reactivity to a mycobacterial antigen mimicking cartilage, HOLOSHITZ et al. (1986) started to screen T lymphocytes collected from patients. One mycobacterial fraction, an acetone precipitate called AP made from the water soluble

Table 2. Mycobacteria and autoimmune arthritis in humans

- Mycobacteria show antigenic mimicry with human cartilage (VAN EDEN et al. 1985b; HOLOSHITZ et al. 1986)
- BCG immunotherapy may cause arthritis (TORISU et al. 1978)
- Rheumatoid arthritis (RA) T cells respond to the AP fraction of mycobacteria (HOLOSHITZ et al. 1986)
- RA T cells respond to the 65K molecule of mycobacteria (RES et al. 1988)
- *M. tuberculosis*-activated RA T cells affect proteoglycan turnover in cartilage explants (WILBRINK et al., manuscript in preparation)
- RA patients have raised IgG antibodies against the 65K molecule of mycobacteria (BAHR et al. 1988)
- HLA-DR4 is a mycobacterial Ir gene (OTTENHOFF et al. 1986)
- RA glycosylation pattern is present in tuberculosis and rat adjuvant arthritis (RADEMACHER et al. in press; COHEN, unpublished data)

fraction of whole mycobacteria, was selected for this purpose. AP is a mycobacterial subfraction apparently enriched for the mimicking epitope, as defined by a superior stimulation of the A2b clone, before the 65K molecule had been identified. Upon screening, the patients showed prominent responses specifically against AP. At the start of the disease, which means up to 1 year after the diagnosis has been made, reactivity was confined to T cells obtained from joint synovial fluids. Later on, reactivity was found in peripheral blood. In normal controls or individuals with degenerative joint disease, no such reactivity was found. These findings indicate that in rheumatoid arthritis specific T-cell reactivity does exist against mycobacterial antigens and cross-reacts with cartilage proteoglycans. That this specific T-cell reactivity is found at the beginning of the disease only in affected joints suggests its relatedness to the pathogenesis of the disease.

Similar findings, which supported the same, were described by RES et al. (1989) from the Leyden University Hospital after they stimulated T cells isolated from patient synovial fluids with our 65K mycobacterial protein. Marked responses were obtained in patients' lymphocytes only when the joint had been clinically affected for less than about 3 years. Nevertheless, we cannot exclude the possibility that, because of the mimicry, the antimycobacterial reactivity is the consequence, not the cause, of the joint's inflammatory process.

As an alternative approach, Wilbrink and coworkers from the University Hospital in Utrecht have started to stimulate peripheral blood lymphocytes obtained from patients with rheumatoid arthritis by using whole *M. tuberculosis* in the presence of explants of human cartilage. It was observed that proteoglycan turnover in the explants was significantly enhanced when the added lymphocytes of patients were stimulated with *M. tuberculosis* compared with stimulation by concanavalin A. This was not seen in lymphocytes obtained from healthy control individuals. Furthermore, the enhanced proteoglycan turnover was not paralleled by enhanced proliferative responses of the donor lymphocytes (WILBRINK et al., manuscript in preparation). In other words, T cells of rheumatoid arthritis patients show unique activities upon stimulation with mycobacterial antigen also at the level of effector mechanisms directly influencing cartilage integrity. Whether the effects on the cartilage are due directly to enzyme or lymphokine release by activated T cells or due indirectly to T-cell regulatory effects on other effector cells is, however, unknown at present.

Furthermore, we have screened sera from arthritis patients for the presence of antibodies directed against the mycobacterial 65K molecule. In both patients and healthy controls antibodies are found, compatible with the omnipresence of the molecule in bacteria. But high titers of such antibodies are also present in patients with a history of reactive forms of arthritis (WAUBEN et al., manuscript in preparation). Furthermore, Bahr and coworkers have found raised levels of IgG antibodies against the 65K molecule in patients with rheumatoid arthritis (BAHR et al. 1988). Another observation, possibly relevant to the present discussion, was made by OTTENHOFF and coworkers (1986). In a study on the immunogenetics of antimycobacterial skin test responses in Spanish patients with leprosy, HLA-DR4-positive individuals showed enhanced DTH reactivity upon *M. tuberculosis* tuberculin skin testing, as compared with individuals without HLA-DR4. The phenomenon was not related to a certain form of leprosy and was not observed with tuberculin made from additionally tested atypical mycobacteria. Thus, in individuals extensively exposed to mycobacteria, the presence of HLA-DR4 correlates with a marked T-cell reactivity specific for an *M. tuberculosis* antigen. HLA-DR4 has not been associated with leprosy, but it is a well-known marker for genetic predisposition to developing rheumatoid arthritis in almost all populations studied (GRENNAN and DYER, 1988). The new findings were interpreted to indicate that HLA-DR4 is an immune response gene with specificity for *M. tuberculosis*. It is just this aspect of HLA-DR4 that might explain its association with rheumatoid arthritis. So, also at the genetic level there is evidence for a connection between recognition of certain mycobacterial antigens and rheumatoid arthritis. Recently, the same connection seemed implicated in unexplained changes in glycosylation of immunoglobulins and possibly glycoproteins in general. Rheumatoid arthritis patients exhibit a characteristic change in the composition of the oligosaccharide molecules present on their serum IgG (PAREKH et al. 1985). Many other autoimmune diseases and infectious diseases now have been screened for this specific change in glycosylation, and the peculiar finding is that, besides rheumatoid arthritis, this phenomenon is almost invariably present in tuberculosis and Crohn's disease. No other diseases, including systemic lupus erythematosus, sarcoidosis, scleroderma, ulcerative colitis, or leprosy, show signs of a similar change in glycosylation (RADEMACHER et al. in press). Moreover, first experiments have indicated that a comparable glycosylation change can be induced in rats by immunization with mycobacteria leading to AA (Cohen, unpublished data). Although the significance of the glycosylation changes for the pathogenesis of rheumatoid arthritis is unknown as yet, this phenomenon again seems compatible with the hypothesis that rheumatoid arthritis is connected to immunological contact with mycobacteria. Despite suggestive evidence from various sites for this connection, clinically there is no indication of a significant association between mycobacterial disease and rheumatoid arthritis. Furthermore, there is no evidence that rheumatoid arthritis has a high prevalence in areas with much mycobacterial disease or that the incidence of rheumatoid arthritis is declining together with a decline in mycobacterial disease or the abolishment of BCG vaccination. Perhaps we have to conclude that, in case mycobacterial antigens indeed are responsible for the triggering of autoimmune arthritis, immunization can be achieved through subclinical exposure to possibly atypical mycobacteria. Alternatively, since the 65K heat-shock protein homologue is present in virtually all bacterial species, mycobacterium may be just one of the possible candidates among many other organisms.

6 Mimicry and the Restoration of Self-Tolerance

As discussed above, Lewis rat T cells are prone to react to a cartilage-mimicking epitope when confronted with mycobacteria. The inevitable consequence of this seems to be loss of self-tolerance ensuing in autoimmune arthritis. Interestingly, however, affected rats do recover spontaneously after a period varying from 2 weeks to some months (see Fig. 3). Apparently, even individuals that tend to focus their immune response on dangerously self-mimicking epitopes, when exposed to some exogenous antigen (Fig. 5), have the capacity of regaining the proper balance in the network and therefore their self-tolerance. This is likely to be effected by the activity of regulatory T cells, which are capable of controlling the activity of the cells that respond to the "incorrect" epitope. A cell possibly interacting in this network of regulatory cells in AA was isolated by further subcloning of the original arthritogenic A2 line. Amongst several subclones lacking apparent *in vivo* functional activity, one subclone was obtained that had protective effects *in vivo*. Intravenous inoculation of this clone, called A2c, never produced arthritis, and subsequent attempts to induce active arthritis with mycobacteria in such rats invariably failed. Furthermore, inoculation of A2c in rats with AA accelerated remission of the disease (COHEN et al. 1985). We were apparently dealing with a T cell clone that had the potential of preventing or therapeutically aborting the autoimmune process. Possibly, this clone was a member of those cells responsible for the vaccination effects of A2 in nonirradiated recipients.

Experiments were undertaken to investigate the mechanism of the protective activity of A2c. In cocultures, A2c had no detectable direct effect on A2 or A2b. However, draining lymph node cells obtained from rats primed with A2c were found to exert a suppressive effect on the A2b response in the presence of *M. tuberculosis*. This effect was not seen with lymph node cells taken after priming with irrelevant control clones,

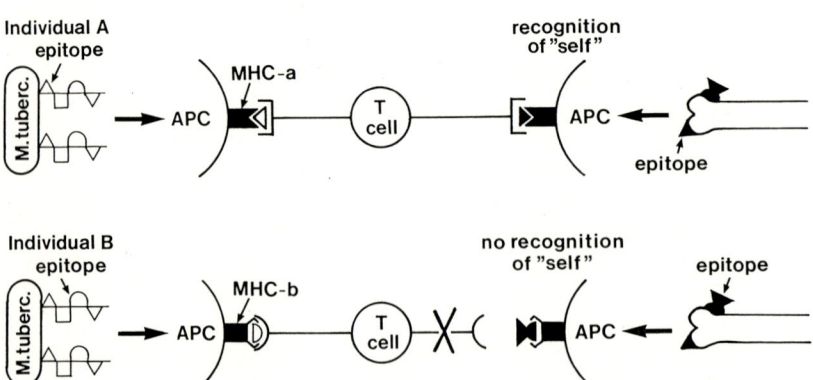

Fig. 5. Schematic representation of possible events determining the MHC influence on the recognition of cartilage-mimicking epitopes in mycobacteria. Two individuals carrying different MHC types are confronted with *M. tuberculosis*. *Individual A* tends to select a dangerously cross-reactive epitope for presentation to the T cell system and might develop autoimmune arthritis. *Individual B* carries an MHC molecule that selects another, more innocent epitope for T-cell recognition, and no dangerous cross-reactivity is generated. Thus individual A is susceptible to autoimmune arthritis induced by *M. tuberculosis*, whereas individual B is not

and it was specific for A2b responses. A suppressive influence on an unrelated cell line with a different antigenic specificity occurred only when the A2b clone and its antigen, *M. tuberculosis*, were present simultaneously in the culture. These findings indicated that A2c could recruit *in vivo* cell populations capable of suppressing lymphocyte proliferation, depending on the specific presence of both the A2b clone and *M. tuberculosis*. Apparently, depending on the specific presence of certain clones together with their target antigens, A2c can initiate a sequence of events that leads to non specific suppression of T-cell activation, probably mediated by cytokines active at short distances. This may explain how a single T cell clone, A2c, with a single receptor specificity, can induce protection against local joint inflammations, which are likely to involve activation of cells with multiple specificities. It may also explain how this suppression is directed to sites where a relevant target antigen is present, for instance in the joint in the case of arthritis. A process involving target antigens at a different site, such as that of experimental allergic encephalomyelitis (EAE) lesions, was not influenced by immunization with A2c. Since A2c has, like A2 and A2b, the helper/DTH phenotype, A2c was termed a suppressor inducer cell (COHEN 1986). This would imply that a single suppressor inducer clone may initiate a course of events that leads to specific resistance against autoimmune arthritis.

Since the A2c clone was present, like the arthritogenic A2b, in the original A2 line, it is possible that at the time of immunization with *M. tuberculosis* both the "aggressive" and the "protective" T-cell elements are triggered. In other words, this suggests that regulation forms an integral part of the response upon antigenic stimulation. Alternatively, perhaps A2c arose due to *in vitro* cellular dedifferentiation. When we screened A2c, like A2b, for its fine specificity, it recognized the same 180–188 epitope of the 65K mycobacterial protein (VAN EDEN et al. 1988). So, A2b and A2c probably deploy identical receptor specificities and, therefore, idiotypes. Such an idiotypic identity between them might be an explanation for the fact that

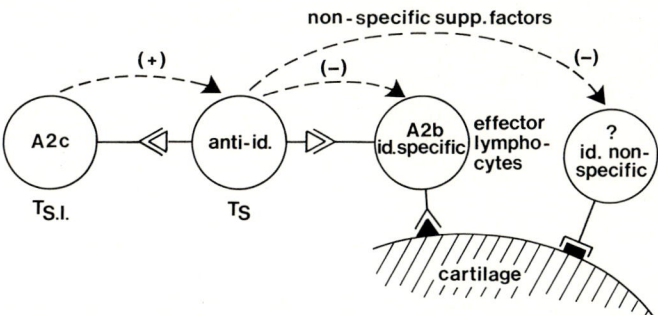

Fig. 6. Immunization with T cell clone *A2c* in Lewis rats leads to specific protection adjuvant arthritis. *In vitro*, lymphocytes from A2c-immunized rats suppressed proliferation of A2b and other irrelevant T cells, depending on the presence of A2b and *M. tuberculosis*. Since A2b and A2c were shown to have identical antigenic specificities, these phenomena can be explained by the interactive role of a suppressive cellular compartment (T_s) that has an anti-idiotypic relationship with both A2b and A2c. T_s apparently can become induced by the encounter with A2c *in vivo*. Upon activation by T cells exposing the right (*A2b*) idiotype, suppressive activity, probably mediated by soluble factors, is generated. This suppression also affects *idiotype-nonspecific* T cells in the close vicinity. This hypothetical course of events might explain the protective activity of A2c as a single clone against a disease process that possibly involves clones of more than one specificity

A2c-induced regulation was dependent on the presence of A2b (Fig. 6). Moreover, idiotype sharing might also help to explain why receptor cross-linking of A2b endows A2b with protective characteristics, comparable to A2c (LIDER et al. 1987). Receptor aggregation is known to enhance the potency for generating an anti-idiotypic response (COHEN 1986, 1988).

Irrespective of the exact mechanism of the functional difference between A2b and A2c, for the present discussion it seems relevant to conclude from the foregoing that a mimicry epitope may activate, harmless or even protective regulatory cells besides autoreactive cells. This can explain the observation that immunization with the isolated 65K molecule did not induce arthritis but induced protection against AA instead (VAN EDEN et al. 1988). Thus, although *M. tuberculosis* induces arthritis because of a mimicking epitope present on its 65K molecule, immunization with the 65K molecule itself, when separated from the context of whole *M. tuberculosis*, may elicit a protective type of response. So, it seems possible, at least in the case of AA, to manipulate the regulatory cells without causing an imbalance in the network leading to autoimmunity. We have shown that it can be done with the antigen that exhibits the mimicry itself. A form of specific vaccination can be achieved by using, in analogy with the vaccination against infectious diseases, the structure responsible for the induction of the disease. However, whether the findings obtained in the model of AA will apply to arthritis in the clinical situation remains to be seen. In the end, it might well turn out to be easier to raise the level of specific resistance to disease by activating regulatory elements present in the fine-tuned regulatory network of the healthy person, than to interfere with a profound process of chronic dysregulation with irreversible damage in the patients. From the natural course of disease, as exhibited by tuberculoid leprosy and AA, with spontaneous remission, it is apparent that the natural regulatory mechanisms that inhibit the immunopathological response remain essentially intact. Whether that is also the case in rheumatoid arthritis, we do not know. If not in the advanced case, it might still be so at the start of disease. The possibility of exploiting the natural mechanisms that normally form a hedge against autoimmunity seems most attractive. Mindful of mimicry as a threat to self-integrity, we should realize that the challenge of present day immunology cannot be to change the outside world. The real challenge, however, will be to train the immune system of the unfortunate patient to resume control over its pathological response to the endogenous self.

References

Bahr GM, Rook GAW, Al-Saffar M, Van Embden JDA, Stanford JL, Behbehani K (1989) An analysis of antibody levels to mycobacteria in relation to HLA type: evidence for non-HLA-linked high levels of antibody to the 65kD heat shock protein of *M. tuberculosis* in rheumatoid arthritis. Clin Exp Immunol 74: 211–215

Battisto JR, Smith RN, Bechman K, Sternlight M, Welles WL (1982) Susceptibility to adjuvant arthritis in DA and F344 rats. A dominant trait controlled by an autosomal gene locus linked to the major histocompatibility complex. Arthritis Rheum 25: 1194–1200

Ben-Nun A, Cohen IR (1982) Experimental autoimmune encephalomyelitis (EAE) mediated by T cell lines: process of selection of lines and characterisation of the cells. J Immunol 129: 303–308
Caplan AI (1984) Cartilage. Sci Am 251: 82–90
Cohen IR (1986) Regulation of autoimmune disease: physiological and therapeutic. Immunol Rev 94: 5–21
Cohen IR (1988) The self, the world, and autoimmunity. Sci Am 255: 34–42
Cohen IR, Holoshitz J, Van Eden W, Frenkel A (1985) Lines of T lymphocytes illuminate pathogenesis and affect therapy of experimental arthritis. Arthritis Rheum 28: 841–845
Crawford CL, Evans DML, Evans EM (1974) Experimental allergic neuritis induced by sensory nerve myelin may provide a model for nonlepromatous leprosy. Nature 251: 223–225
De Vries RRP, Lai-A-Fat RFM, Nyenhuis LE, Van Rood JJ (1976) HLA-linked genetic control of host response to *Mycobacterium leprae*. Lancet ii: 1328–1339
Emmrich F, Thole J, Van Embden J, Kaufmann SHE (1986) A recombinant 64 kilodalton protein of *Mycobacterium bovis* BCG specifically stimulates human T4 clones reactive to mycobacterial antigens. J Exp Med 163: 1024–1029
Fox R, Sportsman R, Rhodes G, Luka J, Pearson G, Vaughan J (1986) Rheumatoid arthritis synovial membrane contains a 62000 MW protein that shares an antigenic epitope with the Epstein-Barr virus encoded associated nuclear antigen. J Clin Invest 77: 1539–1547
Grennan DM, Dyer PA (1988) Immunogenetics and rheumatoid arthritis. Immunol Today 9: 33–34
Guillet JG, Lai MZ, Briner TJ, Buus S, Sette A, Grey HM, Smith JA, Gefter ML (1987) Immunological self, non-self discrimination. Science 253: 865–870
Holoshitz J, Naparstek Y, Ben-Nun A, Cohen IR (1983) Lines of T lymphocytes induce or vaccinate against autoimmune arthritis. Science 219: 56–58
Holoshitz J, Matitiau A, Cohen IR (1984) Arthritis induced in rats by clones of T lymphocytes responsive to mycobacteria but not to collagen type II. J Clin Invest 73: 211–215
Holoshitz J, Klajman A, Drucker I, Lapidot Z, Yaretzky A, Frenkel A, Van Eden W, Cohen IR (1986) T-lymphocytes of rheumatoid arthritis patients show augmented reactivity to a fraction of mycobacteria cross-reactive with cartilage. Lancet ii: 305–309
Kaufmann SHE, Vath U, Thole JER, Van Embden JDA, Emmrich F (1987) Enumeration of T cells reactive with *Mycobacterium tuberculosis* organisms and specific for the recombinant mycobacterial 64 kD protein. Eur J Immunol 17: 351–357
Lider O, Karin N, Shinitzki M, Cohen IR (1987) Therapeutic vaccination against adjuvant arthritis using autoimmune T cells treated with hydrostatic pressure. Proc Natl Acad Sci USA 84: 4577–4580
Matzinger P, Zamoyska R, Waldmann H (1984) Self tolerance is H-2 restricted. Nature 308: 738–741
McMullin TW, Holberg RL (1988) A highly evolutionarily conserved mitochondrial protein is structurally related to the protein encoded by the *Escherichia coli* groEL gene. Mol Cell Biol 8: 1
Mehra V, Convit J, Rubinstein A, Bloom BR (1982) Activated suppressor T cells in leprosy. J Immunol 129: 1946–1951
Mehra V, Brennan PJ, Rada E, Convit J, Bloom B (1984) Lymphocyte suppression in leprosy induced by unique *M. leprae* glycolipid. Nature 308: 194–196
Neame TJ, Chrestner JE, Baker JR (1986) The primary structure of link-protein from rat chondrosarcoma proteoglycan aggregate. J Biol Chem 261: 3519–3535
Ottenhoff THM, Gonzalez NM, De Vries RRP, Convit G, Van Rood JJ (1984) Association of HLA specificity LB-E12 (MB1, DC1, MT1) with lepromatous leprosy in a Venezuelan population. Tissue Antigens 24: 25–29
Ottenhoff THM, Torres P, De Las Aguas JT, Fernandez R, Van Eden W, De Vries RRP, Stanford JL (1986) Evidence for an HLA-DR4-associated immune response gene for *Mycobacterium tuberculosis*: a clue to the pathogenesis of rheumatoid arthritis. Lancet ii: 310–313
Parekh RB, Dwek RA, Sutton BJ, Fernandes DL, Leung A, Stanworth DR, Rademacher TW, Mizuochi T, Taniguchi T, Matsuta K, Takeuchi F, Nagano Y, Miyamoto T, Kobata A (1985) Association of rheumatoid arthritis and primary osteoarthritis with changes in the glycosylation pattern of total serum IgG. Nature 316: 452–457

Pearson CM (1956) Development of arthritis, periarthritis and periostitis in rats given adjuvant. Proc Soc Exp Biol Med 91: 95–101

Pearson CM, Wood FD (1964) Studies of polyarthritis and other lesions induced in rats by injection of mycobacterial adjuvant: 1. General clinical and pathologic characteristics and some modifying factors. Arthritis Rheum 2: 440–459

Rammensee HG, Bevan MJ (1984) Evidence from in vitro studies that tolerance to self antigens is MHC-restricted. Nature 308: 741–744

Res PCM, Schaar CG, Breedveld FC, Van Eden W, Van Embden JDA, Cohen IR, De Vries RRP (1988) Synovial fluid T cell reactivity against the 65kD heat-shock protein of mycobacteria in early onset of chronic arthritis. Lancet ii: 478–481

Ridley DS, Jopling WH (1966) Classification of leprosy according to immunity. A five group system. Int J Lepr 34: 225–273

Sercarz E (1987) Ir-gene regulation: Past failures to present cogent mechanisms and to delete diverting oversimplifications — A commentary. J Mol Cell Immunol (1987) 3: 277–278

Shinnick TM, Vodkin MH, Williams JC (1988) The *Mycobacterium tuberculosis* 65kD antigen is a heat-shock protein which corresponds to common antigen and to the *Escherichia coli* GroEL protein. Infect Immun 56: 446–451

Thole JER, Dauwerse HG, Das PK, Groothuis DG, Schouls LM, Van Embden JDA (1985) Cloning of the *Mycobacterium bovis* BCG DNA and expression of antigens in *Escherichia coli*. Infect Immun 50: 800–806

Thole JER, Keulen W, Kolk A, Groothuis D, Berwald L, Tiesjema R, Van Embden J (1987) Characterisation, sequence determination and immunogenicity of a 64-kilodalton protein of *Mycobacterium bovis* BCG, expressed in *Escherichia coli* K-12. Infect Immun 55: 1466–1475

Thole JER, Hindersson P, De Bruyn J, Cremers F, Van der Zee J, De Cock H, Tommassen J, Van Eden W, Van Embden JDA (1988) Antigenic relatedness of a strongly immunogenic 65kD mycobacterial protein antigen with a similarly sized ubiquitous bacterial common antigen. Microbiol Pathogen 4: 71–83

Torisu M, Miyahara T, Shinohara N, Ohsato K, Sonozaki H (1978) A new side effect of BCG immunotherapy: BCG-induced arthritis in man. Cancer Immunol Immunother 5: 77–83

Van den Broek MF, Van den Berg WB, Arntz OJ, Van de Putte LBA (1988) Reaction of bacterium primed murine T cells to cartilage components: a clue to the pathogenesis of arthritis? Clin Exp Immunol 72: 9–14

Van der Zee R, Van Eden W, Meloen RH, Noordzÿ A, Van Embden JDA (1989) Efficient mapping and characterization of a T cell epitope by the simultaneous synthesis of multiple peptides. Eur J Immunol 19: 43–47

Van Eden W, De Vries RRP, D'Amaro J, Schreuder GMTh, Leiker DL, Van Rood JJ (1982) HLA-DR associated genetic control of the type of leprosy in a population from Surinam. Hum Immunol 4: 343–350

Van Eden W, De Vries RRP, Stanford GL, Rook GAW (1983) HLA-DR3 associated genetic control of response to multiple skin tests with new tuberculins. Clin Exp Immunol 51: 207–292

Van Eden W, De Vries RRP (1984) HLA and leprosy: a reevaluation. Lepr Rev 55: 89–104

Van Eden W, Elferink BG, Hermans J, De Vries RRP, Van Rood JJ (1984a) Role of HLA-class 2 products in proliferative T lymphocyte responses to PPD: evidence for a regulatory influence associated with MB1. Scand J Immunol 20: 503–510

Van Eden W, Elferink BG, De Vries RRP, Leiker DL, Van Rood JJ (1984b) Low T lymphocyte responsiveness to *M. leprae* antigens in association with HLA-DR3. Clin Exp Immunol 55: 140–148

Van Eden W, Gonzalez NM, De Vries RRP, Convit J, Van Rood JJ (1985a) HLA-linked control of predisposition to lepromatous leprosy. J Infect Dis 151: 9–14

Van Eden W, Holoshitz J, Nevo Z, Frenkel A, Klajman A, Cohen IR (1985b) Arthritis induced by a T lymphocyte clone that responds to *Mycobacterium tuberculosis* and to cartilage proteoglycans. Proc Nat Acad Sci USA 82: 5064–5067

Van Eden W, Holoshitz J, Cohen IR (1986) Autoimmunity and the host-parasite relationship. Clinics Immunol Allerg 6: 113–126

Van Eden W, Holoshitz J, Cohen IR (1987) Antigenic mimicry between mycobacteria and cartilage proteoglycans: the model of adjuvant arthritis. In: Cruse TM (ed) Concepts Immunopathol 4. Karger, Basel, pp 144–170

Van Eden W, Thole JER, Van der Zee R, Noordzij A, Van Embden JDA, Hensen EJ, Cohen IR (1988) Cloning of the mycobacterial epitope recognised by T lymphocytes in adjuvant arthritis. Nature 331: 171–173

Van Schooten WCA, Ottenhoff THM, Klatser PR, Thole JER, De Vries RRP, Kolk AHJ (1989) T cell epitopes on the 36 K and 65 K *M. leprae* antigens defined by human T cell clones. (to be published)

Young DB, Ivanyi J, Cox JH, Lamb JR (1987) The 65kDa antigen of mycobacteria — a common bacterial protein? Immunol Today 8: 215–219

Klebsiella Pneumoniae and HLA B27-Associated Diseases of Reiter's Syndrome and Ankylosing Spondylitis

P. L. Schwimmbeck[1,2] and M. B. A. Oldstone[1]

1   Introduction   45
2   Pathogenesis of Spondyloarthropathies   46
3   Homologies Between Microbes and HLA B27   47
4   Cross-Reactivity Between *Klebsiella pneumoniae* and HLA B27 on a molecular Level: Molecular Mimicry   48
4.1   Identification of Homologies   48
4.2   Immunological Cross-Reactivity   49
4.3   Expression of the Epitope Shared by HLA B27 and *Klebsiella pneumoniae* Nitrogenase   51
4.4   Serological Findings in Patients with Ankylosing Spondylitis and Reiter's Syndrome   51
5   Concluding Remarks   54
References   54

1 Introduction

Ankylosing spondylitis (AS) and Reiter's syndrome (RS) are two major nonrheumatoid arthritic diseases of unknown etiology. They are good examples of a disease category characterized by the presence of certain genetic features of the patient combined with a particular exogenous antigen. Both AS and RS show a familial pattern of distribution, but not only genetic factors are involved in the pathogenesis of AS, since monozygotic twins differ considerably in the incidence of disease (Eastmond and Woodrow 1977). Although exogenous factors influence the manifestation of disease, infections by any of a wide variety of bacteria have been identified preceding the onset of AS. Most often cited among these are *Klebsiella*, *Shigella*, *Salmonella*, *Yersinia*, and *Escherichia coli* (Sheldon 1985). Another marked feature of both AS and RS is the high correlation with HLA type B27. Several studies have shown that more than 90% of Caucasian patients with AS and more than 80% of those with RS are HLA B27-positive, whereas only about 9% of the normal Caucasian population has this HLA haplotype (Brewerton et al. 1973; Schlosstein et al.

---

[1] Dept. of Immunology, Research Institute of Scripps Clinic, 10666 No. Torrey Pines Road, La Jolla, CA 92037, USA
[2] present address: Dept. of Internal Medicine/Cardiology, University of Düsseldorf, Moorenstrasse 5, 4000 Düsseldorf 1, FRG

1973). Therefore, the relative risk factor for HLA B27-positive individuals of acquiring the disease is 87 for AS and 37 for RS, as compared with the normal Caucasian population (SVEJGAARD et al. 1983).

2 Pathogenesis of Spondyloarthropathies

Among the many hypotheses attempting to explain the involvement of HLA B27 and microbes in the pathogenesis of AS, two major groups can be separated (KEAT 1986). The first postulates that genes genetically linked to HLA B27, but not identical with it, encode an immune response predisposing for AS. Thus, a defect in the handling of a microorganism, or special classes of them, would render a genetically predisposed individual susceptible to AS when infected by the microbe. Such immune-responsiveness genes are known from studies in mice in which they govern the immune response to viral antigens, but no equivalent genes are known in humans (MC DEVITT 1984). In this light, quite extensive studies of the immune status of HLA B27-positive patients with or without AS or RK were carried out (reviewed in YU 1988). By using monoclonal antibodies, the levels of T lymphocytes in their peripheral blood were measured and found to be normal (VEYS et al. 1983). Further studies examining the in vitro lymphocyte responses to a variety if mitogens or antigens reported somewhat lowered responses (SHELDON 1985; GOEBEL et al. 1982; FAN et al. 1977), but there was a considerable overlap between the patients and control subjects. The study of IgA levels in patients with *Yersinia* infections, only half of whom developed arthritis, showed no difference in the total amount if IgA (GRANFORS et al. 1978). Finally, when circulating immune complexes were studied, slightly elevated levels were found in patients with AS and RS, but still much lower than the levels seen in patients with rheumatoid arthritis or systemic lupus erythematosus (ROSENBAUM et al. 1981; MANICOURT and ORLOFF 1981). In summary, the data mentioned above show that it is rather unlikely that a generalized immune abnormality is present in patients with AS or RS, despite the lack of agreement among the various reports.

The alternative hypothesis proposes that the HLA B27 antigen is directly involved in the tissue-damaging process, such as by functioning as a recepter molecule for immune mediators of the inflammatory process (e.g., lymphokines or monokines), thus causing an inappropriate immune response, or by binding of "modifying factor(s)" present in culture filtrates of bacteria associated with AS (GECZY et al. 1980). As shown by Geczy and coworkers, it is possible to modify the antigenic appearance of B27-positive cells from nonspondylitic individuals by incubation with a modifying factor, produced by a wide array of bacteria like *Salmonella*, *Shigella*, and *E. coli* (PRENDERGAST et al. 1983). After such incubation, the cells resemble those from B27-positive spondylitic patients and could, subsequently, be lyzed by a serum raised against *Klebsiella* (SEAGER et al. 1979). However, only lymphocytes from B27-positive individuals could be modified, thus suggesting that the B27 molecule itself was the receptor for the modifying factor(s). Studies from the same group indicate that the modifying factor is a component of the outer membrane of bacteria, having a molecular weight of about 26K–30K (DRUERY et al. 1980). It is heat labile and sensitive to treatment with neuraminidase, although stable to treatment with trypsin or chymotrypsin. Since the modifying capacity is present in only about 8% of cultures of *Klebsiella* and is shared by a variety of bacteria, it was hypothesized that the capacity to produce it is transmitted by a plasmid between microbes, quite similar to the transfer of resistance to antibiotics (CAMERON et al. 1983). The studies of the production and actions of modifying factor, however, have not yet been confirmed by other investigators (KINSELLA et al. 1983; CAMERON et al. 1987).

Alternatively, the HLA B27 molecule may share antigenic determinats, either linear or conformational, with microbes associated with AS and thus react as the autoantigen for an immune response initiated by a microbe associated with AS during an infection. Such an autoimmune process, as shown in a variety of diseases in this volume, could be self-sustaining by the release of the autoantigen from destroyed cells (OLDSTONE and NOTKINS 1986). When looking for candidates for the initiation of a cross-reacting autoimmune response, several bacteria were found either to have caused an infection in spondylitic patients or to remain present in their bowels. Earlier studies along these lines reported the isolation of *K. pneumoniae* in a high percentage of patients with active AS but in only a minority of healthy controls (EBRINGER et al. 1978). Additional studies correlated the presence of *Klebsiella* in the bowels of patients with AS with an exacerbation of this disease and showed increased levels of IgA antibodies against *Klebsiella* in the sera of these patients (TRULL et al. 1983).

3 Homologies Between Microbes and HLA B27

In studying molecular mimicry between HLA B27 and microbes, a variety of cross-reacting determinants found in a wide array of bacteria were identified (Table 1). The initial reports involved antisera raised against human B27-positive lymphocytes. Such antibodies were shown to bind to *Klebsiella aerogenes* antigens in a gel diffusion assay, and antisera raised against that microbe specifically lyzed B27-positive lymphocytes of patients with AS in a complement-dependent assay (EBRINGER 1983; SEAGER et al. 1979). Further studies suggest that the cross-reactivity between HLA B27 and *Klebsiella* is not only restricted to that microbe but is also shared by clinical isolates of *Salmonella, Shigella, E. coli,* and *Campylobacter*, which absorbs the specific cytotoxicity of *Klebsiella*-specific sera to HLA B27-positive lymphocytes of patients with AS (PRENDERGAST et al. 1983). The molecular

Table 1. Cross-reactivities between microbes and HLA B27

Microorganism, protein	Homologous cell protein	Ref.
Klebsiella	HLA B27 + cells	EBRINGER 1983
Salmonella, Shigella,. *E. coli, Campylobacter*	B27 + cells of patients ankylosing spondylitis	PRENDERGAST et al. 1983
Klebsiella, Shigella, Yersinia	HLA B27	VAN BOHEMEN et al. 1984
Yersinia enterocolotica	HLA B27	OGASAWARA et al. 1985
Yersinia pseudotuberculosis	HLA B27	CHEN et al. 1987
Klebsiella pneumoniae nitrogenase	HLA B27	SCHWIMMBECK et al. 1987

weight of such cross-reacting proteins was identified in studies using monoclonal antibodies raised against HLA B27. Using the Western blot procedure, the cross-reacting antigens were shown to be a 16K protein of *Yersinia entercolitis*, a 21K and a 43K protein of *K. pneumoniae*, and a 20K protein of *Shigella flexneri* (VAN BOHEMEN et al. 1984). Additional cross-reacting antigens were identified in *Y. enterocolitica* and *Y. pseudotuberculosis* by using monoclonal antibodies against the microbe (OGASAWARA et al. 1985). A further cross-reacting antigenic determinant was detected on *Y. pseudotuberculosis* and HLA B27-positive cells (CHEN et al. 1987). Finally, our own results, as shown in the following paragraphs, identified an homology of six consecutive amino acids between HLA B27.1 and *K. pneumoniae* nitrogenase, a 32K protein (SCHWIMMBECK et al. 1987; SCHWIMMBECK and OLDSTONE 1988).

4 Cross-Reactivity Between *Klebsiella pneumoniae* and HLA B27 on a molecular Level: Molecular Mimicry

4.1 Identification of Homologies

Molecular mimicry, i.e., the sharing of cross-reacting antigenic determinants, may be due either to conformational or linear antigenic determinants. These antigenic determinats can be detected by cross-reacting antibodies or, especially with linear determinants, by the screening of databases for sequence homologies shared between the putative autoantigen and microbes (e.g., bacteria, viruses, or parasites) associated with the disease (OLDSTONE and NOTKINS 1986). In our studies, we wanted to examine homologies between HLA B27 and microbes associated with AS and RS. Therefore, using the recently determined sequence of HLA B27 (APARICIO et al. 1985; SZOETS et al. 1986), we made a computer search for homologies to the sequences listed in the Dayhoff Data Base, which contains more than 3400 sequences with more than 770 000 residues in the version 5.0 (BARKER et al. 1985). Using the computer program "search" (ORCUTT et al. 1982) and the "unitary matrix" for scaling (ORCUTT and DANHOFF 1982), we found homologies of up to six consecutive amino acids (Table 2). The best homology consisted of six consecutive amino acids shared between HLA B27.1, residues 72–77, and *K. pneumoniae* nitrogenase (SUNDARESAN and AUSUBEL 1981), residues 188–193 (QTDRED), a microbe often clinically associated with AS and RS.

Table 2. Homologies between HLA B27 and microbes

HLA B27.1, res. 72–79:	AQTDREDL
Klebsiella pneumoniae nitrogenase, residues 187–194:	RQTDREDE
HLA B27.1, residues 76–83:	EDLRTLLR
EBV, BLRF3 protein, residues 99–106:	LDLRTLLQ
HLA B27.1, residues 44–51:	REEPRAPW
EBV, BBLF4 protein, residues 2–19:	MEEPRAPE

```
KLEBSIELLA PNEUM. NITROGENASE, 185-194:   N S R Q T D R E D E
HLA B27.1, 69-78:                         A K A Q T D R E D L

HLA B27.2, 69-78:   A K A Q T D R E N L
HLA B27.3, 69-78:   A K A Q T D R E S L
HLA B40, 69-78:     T N T Q T Y R E S L
HLA B7, 69-78:      A Q A Q T D R E S L
HLA A2, 69-78:      A H S Q T H R V D L
HLA A3, 69-78:      A Q S Q T D R V D L
HLA A28, 69-78:     A Q S Q T D R V D L
```

```
                  20              40              60              80              100
HLA-B27.1  GSHSMRYFHTSVSRPGRGEPRFITVGYVDDTLFVRFDSDAASPREEPRAPWIEQEGPEYWDRETQICKAKAQTDREDLRTLLRYYNQSEAGSHTLQNMYG
HLA-B27.2  ----------------------------------------------------------------------------N--IA------------------
HLA-B27.3  -------------------------------------------------------------------------------S-------------------
HLA-B40    ----------AM------------------------T---K-------------------S-TNT--Y--S--N-RG-------------R---

                  120             140             160             180             200
HLA-B27.1  CDVGPDGRLLRGYHQDAYDGKDYIALNEDLSSWTAADTAAQITQRKWEAARVAEQLRAYLEGECVEWLRRYLENGKETLQRADPPKTHVTHHPISDHEAT
HLA-B27.2  ----------------------------------------------------------------------------------------------------
HLA-B27.3  -----------------------------------------------E---------------------------------------------------
HLA-B40    -----------HN-Y-------------R----------   ---L------------------------DK-E--------------------

                  220             240             260
HLA-B27.1  LRCWALGFYPAEITLTWQRDGEDQTQDTELVETRPAGDRTFEKWAAVVVPSGEEQRYTCHVQHEGLPKPLT
HLA-B27.2  ----------------------------------------------------------------------
HLA-B27.3  ----------------------------------------------------------------------
HLA-B40    ---------------------------------Q------------------------------------
```

Fig. 1. Homology between HLA B27 and *Klebsiella pneumoniae* nitrogenase with the homologous amino acids placed in a *box*. The *middle panel* shows the corresponding amino acid sequences of other HLA class I molecules. In the *lower part*, the entire amino acid sequences of HLA B27.1, B27.2, B27.3, and B40 are displayed. The amino acids shared with *Klebsiella* are underlined. (Data from SCHWIMMBECK et al. 1987)

These homologous amino acids are located in the hypervariable region of HLA class I molecules. HLA B27.2 and B27.3, as well as other class I molecules including B40, B7, A2, A3, and A28, differ in at least one or more amino acids in this area (Fig. 1). We also found hexamers shared between HLA B27 and hypothetical proteins of Epstein-Barr virus (EBV) (BAER et al. 1984) (Table 2). However while patients with AS and RS contained antibodies that cross-reacted with *K. pneumoniae* residues 187–194, such patients did not show antibody titers against the EBV regions.

4.2 Immunological Cross-Reactivity

Since *K. pneumoniae* is a pathogen often associated with AS and RS, we decided to study the homology to HLA B27.1 in more detail. One factor influencing immunological cross-reactivity is the secondary structure of a region which is greatly influenced by the local hydrophilicity (WILSON et al. 1984). Therefore, we plotted the hydrophilicity of the shared amino acids (KYTE and DOOLITTLE 1982). As shown in Fig. 2, both sequences show a quite similar hydrophilicity pattern and, therefore, are likely to be accessible to a cross-reacting immune response. Encouraged by these results, we synthesized peptides derived from both HLA B27 and *K. pneumoniae* nitrogenase

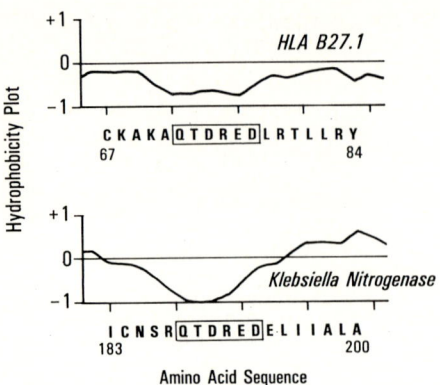

Fig. 2. Hydrophobicity plot of the homology shared between HLA B27.1 and *Klebsiella pneumoniae* nitrogenase. Homologous residues are placed in a *box*. The mean of the hydrophobicity indices over a window of nine amino acids is shown. Hydrophilic domains are displayed by *negative values*, whereas *positive values* indicate hydrophobic stretches. Data shows that the shared homologous regions are hydrophilic.

(Table 3) and raised antibodies to them in both rabbits and rats. The resulting antibodies were affinity purified by using the appropriate peptide coupled to a thiopropyl column, thereby avoiding nonspecific binding. When such antibodies were tested in an ELISA system, they reacted not only with the peptide used for immunization but also with the homolgous peptide containing the shared six amino acids (Table 3). In contrast, the antibodies were not reactive with a peptide derived from another region of HLA B27. Using shorter peptides derived from HLA B27 demonstrated that the cross-reacting antigenic determinant was identical with the amino acids shared between HLA B27 and *K. pneumoniae* nitrogenase, since shorter peptides representing the homologous residues reacted with the antibodies. A single amino acid change, as found in HLA B27.2 (D → N), reduced the reactivity significantly, or as found in HLA B27.3 (D → S), abolished it. These results show that the linear amino acid homology between HLA B27 and *K. pneumoniae* nitrogenase translates into immunological cross-reactivity and that the cross-reacting antigenic determinant is identical with the shared amino acids. In additional experiments done in collaboration

Table 3. Homologies and cross-reactivities between HLA and *Klebsiella pneumoniae* nitrogenase

Peptide	Origin	Location	Amino acid sequence	Antibody reactivity
31	HLA B27.1	Res. 69–84	AKA<u>QTDRED</u>LRTLLRY	0.873
45	*Kleb.* NITR	Res. 185–196+	NSR<u>QTDRED</u>ELIGGC	0.192
43	HLA B27.1	Res. 67–84	CKAKA<u>QTDRED</u>LRTLLRY	1.133
46	HLA B27.1	Res. 68–83+	KAKA<u>QTDRED</u>LRTLLRGGC	0.546
40	HLA B27.1	Res. 69–78	AKA<u>QTDRED</u>L	0.523
41	HLA B27.2	Res. 69–78	AKA<u>QTDREN</u>L	0.117
42	HLA B27.3	Res. 69–78	AKA<u>QTDRES</u>L	0.014
30	HLA B27.1	Res. 44–53	REEPRAPWI	0.000

The homologous amino acids are underlined. Specific antibodies were raised in rabbits against peptide 31 and affinity purified using peptide 43 immobilized on a thiopropyl column. The binding was measured by ELISA at 492 nm. The antibodies were used at a dilution of 1:1600. The peptides marked with „+" are extended at the C-terminal with G—G—C for coupling purposes.

with D. Yu and his colleagues at the University of California, Los Angeles, we could demonstrate that cross-reactivity was also present between the native proteins, i.e., HLA B27, and *K. pneumoniae* nitrogenase (KONO et al. 1988). Using peptide-specific antibodies we showed that the antibodies had a specific cytolytic effect on HLA B27-positive cells and bound to the isolated nitrogenase of *Klebsiella* by Western blot.

4.3 Expression of the Epitope Shared by HLA B27 and *Klebsiella pneumoniae* Nitrogenase

A prerequisite for the concept of molecular mimicry is that the shared epitope is expressed on the affected tissues, e.g., the mucous membranes of joints, and is accessible for a cross-reacting immune response. We addressed this question of accessibility in collaboration with R. C. Williams and his colleagues at the University of New Mexico (HUSBY et al. 1988). For this study, synovial tissues from HLA B27-positive and -negative patients with different arthritic diseases were obtained. When appropriate sections of the synovial tissue were stained with antibodies raised to peptides derived from either HLA B27 or *K. pneumoniae* nitrogenase, both containing the homologous amino acids, strong staining of the synovial lining cells and the endothelial cells in the involved joints of HLA B27-positive patients with AS or RS was observed. In contrast, the same antibodies showed no reactivity with tissues from patients with a variety of arthritic diseases (Table 4). The expression of HLA B27 was markedly enhanced in patients with AS as compared with skin biopsies of healthy HLA B27-positive controls, but it is not known if and how this is influenced by other factors, e.g., immune mediators or lymphokines. These experiments clearly demonstrate that patients with AS have an enhanced expression of the epitope shared by HLA B27 and *K. pneumoniae* nitrogenase in the area of disease activity and that it is accessible for a cross-reacting immune response.

4.4 Serological Findings in Patients with Ankylosing Spondylitis and Reiter's Syndrome

Since we found that the linear amino acid homology between HLA B27 and *K. pneumoniae* nitrogenase translates into immunological cross-reactivity and that the epitope is expressed and accessible in synovial tissue, we wanted to test the pathogenic significance of our findings. Therefore, we obtained sera from HLA B27-positive individuals with and without AS and RS and tested for binding to the various peptides. The representative results of 34 sera from HLA B27-positive patients with RS are shown in Fig. 3. Eighteen of the 34 specimens showed significant titers of antibodies against a peptide derived from HLA B27 and homologous with *K. pneumoniae* nitrogenase. As background, we used the mean +2 standard deviations when we tested the sera of 90 non-HLA-matched individuals not suffering from arthritis. In contrast, none of 22 sera from HLA B27-positive but healthy individuals were reactive at a dilution of 1:10. We found quite similar results when we tested these sera for binding to *K. pneuminiae* nitrogenase (Table 5). We subsequently tested the sera of 60 HLA B27-positive patients with AS for binding to either

Table 4. Reactivity of sera raised against peptides derived from HLA B27 and *Klebsiella pneumoniae* nitrogenase with tissue sections of patients with rheumatic diseases as tested with immune peroxidase staining

Disease	No. of patients	B27	Positive staining with	
			HLA B27, residues 67–84 CKAKAQTDREDLRTLLRY	*K. pneumoniae* nitrogenase, residues 184–196 CNSRQTDREDELI
Ankylosing spondylikis	12	+	11	11
Reactive arthritis	2	+	2	2
Reactive arthritis	5	nil	0	0
Juvenile reactive arthritis	1	nil	0	0
Osteoarthritis	1	nil	0	0
Nonspecific synovitis	1	nil	0	0

Table 5. Binding of patients' sera to peptides derived from HLA B27.1 and *Klebsiella pneumoniae* nitrogenase as tested in ELISA

Peptide	Amino acid sequence	HLA B27-positive individuals			Random HLA
		Reiter's syndrome	Ankylosing spondylitis	No arthritis	
HLA B27.1, residues 69–84	AKAQTDREDLRTLLRY	18/34	16/60	0/22	2/90
Klebsiella nitrogenase, residues 184–196	CNSRQTDREDELI	15/34	18/60	0/22	1/90

The reactivity shown here was tested at a dilution of 1:10. The numbers give positive sera over the numbers of total sera tested.

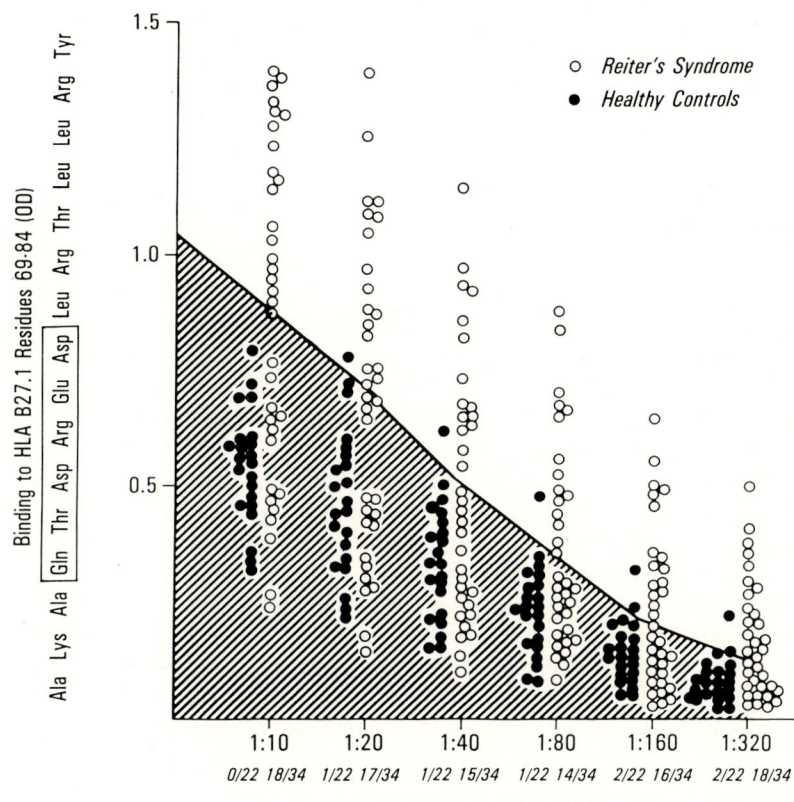

Fig. 3. Binding of sera to HLA B27. The sera were tested for binding to HLA B27.1, residues 69–84, in an ELISA system. The optical density reading at 492 nm is displayed on the *y-axis*, whereas the *x-axis* shows the dilution of the sera and the number of positive sera over the number tested. The background (*shadowed area*) is calculated as the mean plus two standard deviations of 90 non-HLA-matched, nonarthritic controls. The *open circles* represent the values of 34 sera from HLA B27-positive patients with Reiter's syndrome, whereas the *black dots* indicate the values of 22 HLA B27-positive but healthy individuals. (Figure reproduced from SCHWIMMBECK et al. 1987)

peptide, and again we found significant binding for a substantial portion (25%), of patients. Interestingly, when we tested the sera for binding to peptides derived from EBV showing homology to HLA B27, no significant binding occurred (results not shown). We concluded from these results that a substantial portion of patients with AS or RS have antibodies reactive with the epitope shared by HLA B27 and *K. pneumoniae* nitrogenase, which may be induced during an infection with *Klebsiella* containing nitrogenase. An open question is, of course, why not more or all of the patients with AS and RS have antibodies reactive with the epitope we found. This may be due in part to low antibody titer and/or low affinity of the antibodies. Alternatively, only a small portion of bacterial proteins are sequenced and added to the database; therefore, it may very well be possible that other homologies shared between microbes and HLA B27 are missed by our approach. This hypothesis is supported by the recent identification of an epitope shared between HLA B27 and *Yersinia pseudotuberculosis* (Kono et al. 1987) and cross-reactivity between HLA B27 and other microbes such as *Salmonella* and *Shigella* (see Table 1).

5 Concluding Remarks

From our results, we suggest the following scenario for AS and RS. An immune response against a microbial agent like *K. pneumoniae* is raised during an infection. Due to molecular mimicry between HLA B27 and *K. pneumoniae* nitrogenase, the antibodies also recognize the HLA B27 self-determinant whose expression is enhanced in joints. Similarly, under investigation is the role of T-cell mediated injury. T cells recognize the appropriate peptide in the context of the MHC glycoprotein molecules. This results in an autoimmune attack and hence plays an important role in the pathogenesis of AS and RS. By the time the disease becomes overt, the initiating microbe may have been cleared and can no longer be demonstrated directly.

The principles of molecular mimicry, as shown in this volume, have been used for the study of several diseases. Our approach to the study of molecular mimicry is to select a biologically important protein that is supposed to be the target of an autoimmune process. The amino acid sequence is elucidated and compared with all sequences listed in a database by computer search. The best homologies to microbes associated with the disease are identified and tested for their immunological cross-reactivity. The final step is then to evaluate the pathogenic significance of the shared amino acids for autoimmune disease.

Acknowledgement. We would like to thank Nina Bland for her expert technical assistance. For our clinical studies, clinical assessment and sera were kindly provided by Dr. P. Ivanyi (Central Laboratory for Blood Transfusion, Amsterdam, The Netherlands), D. Yu (University of California, Los Angeles), and D. Winfield (University of North Carolina, Chapel Hill). This is publication number 5606-IMM from the Department of Immunology, Scripps Clinic and Research Foundation, La Jolla, CA 92037, USA. This work was supported in part by USPHS grant AI-07007. PLS was the recipient of a grant from the Deutsche Forschungsgemeinschaft (DFG).

References

Aparicio P, Vega MA, Lopez de Castro JA (1985) One allogeneic cytolytic T lymphocyte clone distinguishes three different HLA B27 subtypes: identification of amino acid residues influencing the specificity and avidity of recognition. J Immunol 135: 3074

Baer R, Bankier AT, Biggin MD, Deininger PL, Farrell PJ, Gibson TJ, Hatfull G, Hudson GS, Satchwell SC, Seguin C, Tuffnell PS, Barrell BG (1984) DNA sequence and expression of the B59-8 Epstein-Barr virus genome. Nature 310: 207–211

Barker WC, Hunt LT, George DG, Yeh LS, Chen HR, Blomquist MC, Johnson GC, Seibel-Ross EI, Hong MK, Bair JK, Ledley RS (1985) Protein sequence database of the protein identification resource. National Biomedical Research Foundation, Georgetown University Medical Center, Washington DC

Brewerton DA, Caffrey M, Hart FD, James DCO, Nichols A, Sturrock RD, (1973) Ankylosing spondylitis and HL-A27. Lancet i: 904–907

Cameron FH, Russell PJ, Sullivan J, Geczy AF (1983) Is a *Klebsiella* plasmid involved in the aetiology of ankylosing spondylitis in HLA B27-positive individuals? Mol Immunol 20: 563–566

Cameron FH, Russell PJ, Easter JF, Wakefield D, March L (1987) Failure of *Klebsiella pneumoniae* antibodies to cross-react with peripheral blood mononuclear cells from patients with ankylosing spondylitis. Arthritis Rheum 30: 300–305

Chen J-H, Kono DH, Young A, Park MS, Oldstone MBA, Yu DTY (1987) A *Yersinia pseudotuberculosis* protein which reacts with HLA B27. J Immunol 139: 3003–3011

Druery C, Bashir H, Geczy AF, Alexander K, Edmonds J (1980) Search for *Klebsiella* cell wall components cross-reactive with lymphocytes of B27+ AS+ individuals. Hum Immunol 1: 151–160

Eastmond CJ, Woodrow JC (1977) Discordance for ankylosing spondylitis in monocytic twins. Ann Rheum Dis 36: 360–364

Ebringer A (1983) The cross-tolerance hypothesis, HLA B27 and ankylosing spondylitis. Br J Rheumatol 22 [Suppl 2]: 53–66

Ebringer RW, Cadwell DR, Cowling P, Ebringer A (1978) Sequential studies in ankylosing spondylitis: association of *Klebsiella pneumoniae* with active disease. Ann Rheum Dis 37: 146–151

Fan TP, Clements PJ, Yu DTY, Opelz G, Bluestone R (1977) Lymphocyte abnormalities in ankylosing spondylitis. Ann Rheum Dis 36: 471–473

Geczy AF, Yap J (1982) A survey of isolates of *Klebsiella pneumoniae* which cross-react with HLA B27-associated cell-surface structure on the lymphocytes of patients with ankylosing spondylitis. J Rheumatol 9: 97–100

Geczy AF, Alexander K, Bashir HV (1980) A factor(s) in *Klebsiella* filtrates specifically modifies an HLA B27-associated cell-surface component. Nature 283: 782–784

Goebel KM, Goebel TD, Baier R (1982) Impaired cell-mediated immunity among HLA B27 related rheumatoid variants responding to *Yersinia* antigen. J Clin Lab Immunol 8: 75–81

Granfors K, Viljanen MK, Ahvonen P, Toivanen P (1978) Measurement of IgM and IgA antibodies to *Yersinia* measured by solidphase radioimmunoassay. J Infect Dis 138: 232–240

Granfors K (1979) Measurement of immunoglobulin M (IgM), IgG, and IgA antibodies against *Yersinia enterocolitica* by enzyme-linked immunosorbent assay: persistence of serum antibodies during disease. J Clin Microbiol 9: 336–341

Husby G, Tsuchiya N, Schwimmbeck PL, Keat A, Pahle JA, Oldstone MBA, Williams RC (1989) Cross-reactive epitope with *Klebsiella pneumoniae* nitrogenase in articular tissue of HLA—B27+ patients with ankylosing spondylitis. Arthritis and Rheumatism 32: 437–445

Keat A (1986) Is spondylitis caused by *Klebsiella*? Immunol Today 7: 144–149

Kinsella TD, Fritzler MJ, McNeil DJ (1983) Ankylosing spondylitis: a disease in search of microbes. J Rheumatol 10: 2–4

Kono H, Chen JH, Yu DTY, McLean PA, Schwimmbeck PL, Oldstone MBA (1988) Cross-reactivity between HLA B27 and *Klebsiella pneumoniae* can be demonstrated using antisera generated against synthetic peptides. (to be published)

Kyte J, Doolittle RF (1982) A simple method for displaying the hydropathic character of a protein. J Mol Biol 157: 105–132

Manicourt DH, Orloff S (1981) Immune complexes in polyarthritis after *Salmonella* gastroenteritis. J Rheumatol 8: 613–620

McDevitt HO (1984) Host genes controlling the immune response. In: Notkins AL, Oldstone MBA (ed) Concepts in viral pathogenesis. Springer, Berlin Heidelberg New York, 79–85

Ogasawara M, Kobayashi S, Lahej K, Hill JL, Kono DM, Yu DTY (1985) A heat-modifiable outer membrane protein carries an antigen specific for the species *Yersinia enterocolitica* and *Yersinia pseudotuberculosis*. J Immunol 135: 1430–1436

Oldstone MBA, Notkins AL (1986) Molecular mimicry. In: Notkins AL, Oldstone MBA (eds) Concepts in viral pathogenesis, vol II. Springer, Berlin Heidelberg New York, pp 195–202

Orcutt BC, Dayhoff MO (1982) Scoring matrices. National Biomedical Research Foundation report 820541-08710. National Biomedical Research Foundation, Washington DC

Orcutt BC, Dayhoff MO, Barker WC (1982) User's guide for the database search program search. National Biomedical Research Foundation report 820503-08710. National Biomedical Research

Foundation report 820503-08710. National Biomedical Research Foundation, Georgetown University Medical Center, Washington DC

Prendergast JK, Sullivan JS, Geczy A, Upfold LI, Edmonds JP, Bashir HV, Reiss-Levy E (1983) Possible role of enteric organisms in the pathogenesis of ankylosing spondylitis and other seronegative arthropathies. Infect Immun 41: 935–941

Rosenbaum JT, Theofilopoulos AN, McDevitt HO, Pereira AB, Carson D, Calin A (1981) Presence of circulating immune complexes in Reiter's syndrome and ankylosing spondylitis. Clin Immunol Immunopathol 18: 291–297

Schlosstein L, Terasaki PI, Bluestone R, Pearson CM (1973) High association of an HLA antigen w27 with ankylosing spondylitis. N Engl J Med 288: 704–706

Schwimmbeck PL, Oldstone MBA (1988) Molecular mimicry between HLA B27 and *Klebsiella*: consequences for spondyloarthropathies. Am J Med 85 (6A): 51–53

Schwimmbeck PL, Yu DTY, Oldstone MBA (1987) Autoantibodies to HLA B27 in the sera of patients with ankylosing spondylitis and Reiter's syndrome. J Exp Med 166: 173–181

Seager K, Bashir HV, Geczy AF, Edmonds J, De Vere-Tyndall A (1979) Evidence for a specific B27-associated cell surface marker on lymphocytes of patients with ankylosing spondylitis. Nature 277: 68–70

Sheldon P (1985) Specific cell-mediated response to bacterial antigens and clinical correlations in reactive arthritis, Reiter's syndrome and ankylosing spondylitis. Immunol Rev 86: 5–25

Sundaresan V, Ausubel M (1981) Nucleotide sequence of the gene coding for the nitrogenase iron protein from *Klebsiella pneumonia*. J Biol Chem 256: 2808–2812

Svejgaard A, Platz P, Ryder (1983) HLA and disease 1982 — a survey. Immunol Rev 70: 193–218

Szoets H, Riethmueller G, Weiss E, Meo T (1986) Complete sequence of HLA B27 cDNA identified through the characterization of structural markers unique to the HLA-A, -B, and -C allelic series. Proc Natl Acad Sci USA 83: 1428–1432

Trull AK, Ebringer R, Panayi G, Colthorpe D, James DCO, Ebringer A (1983) IgA antibodies to *Klebsiella pneumoniae* in ankylosing spondylitis. Scand J Rheumatol 12: 249–253

van Bohemen CHG, Grumet FC, Zanen HC (1984) Identification of HLA B27 M1 and M2 cross-reactive antigens in *Klebsiella*, *Shigella*, and *Yersinia*. Immunology 52: 607–610

Veys EM, Verbrugger G, Hermanns P, Mielants H, Van Bruwaene P, DeBrabanter G, DeLandsheere D, Immesoete C (1983) Peripheral blood T lymphocyte subpopulations in HLA B27 related rheumatoid diseases: ankylosing spondylitis and reactive synovitis. J Rheumatol 10: 140–142

Wilson IA, Niman HL, Houghten RA, Cherenson AR, Conolly ML, Lerner RA (1984) The structure of an antigenic determinant in a protein. Cell 37: 767–778

Yu DTY (1989) Experimental studies of infectious agents in reactive arthritis. In: Arnett FC (ed) Infections in the rheumatic diseases. (in press)

Molecular Mimicry and Microorganisms: A Role in the Pathogenesis of Myasthenia Gravis?

M. E. Dieperink and K. Stefansson

1 Introduction 57
2 Sharing of Epitopes by Virus and the Acetylcholine Receptor 59
3 Sharing of Epitopes by the α-Chain of the Acetylcholine Receptor and Gram-Negative Enteric Bacteria 60
4 Antigens on Bacteria Could Cause Myasthenia Gravis Through Sharing of Epitopes by Antibodies Against Bacterial Components and the Acetylcholine Receptor 62
5 Conclusions 63
References 63

1 Introduction

Myasthenia gravis (MG) is a human disease characterized by excessive fatigability and weakness. The disease was originally described by Thomas Willis in 1672 (Herrmann 1967). However, the pathogenesis remained largely a mystery until Patrick and Lindstrom (1973) made a spectacular discovery. They were attempting to raise antibodies against the nicotinic acetylcholine receptor (AchR) by immunizing rabbits with AchR isolated from fish electric organs. Interestingly, the rabbits became sick with a disease that both clinically and electrophysiologically was similar to MG and could be reversed with acetylcholine esterase inhibitors (Grob et al. 1981; Lindstrom 1979). They had developed an animal model for MG. Subsequent work demonstrated that the animal disease could be passively transferred to other animals by antibodies from the actively immunized animals (Lindstrom et al. 1976a), and animals could be made myasthenic by antibodies from the sera of patients with MG. More recently, it has been shown that the experimental disease can be induced with monoclonal antibodies against the α-chain of AchR (Gomes et al. 1981). It was concluded that MG is caused by a humoral immune response against the AchR.

It is of interest here that, when patients are compared with each other, it has proven difficult to find a correlation between titer of AchR-specific antibodies and severity of the disease. However, when longitudinal studies are done on patients, there is for each patient a fair correlation between the clinical status and the AchR-specific antibody titer (Lindstrom et al. 1976b).

Departments of Neurology and Pathology (Neuropathology) and the Committees on Immunology and Neurobiology, The University of Chicago, USA

Assuming that it has been proven beyond reasonable doubt that MG is, at least in part, caused by antibodies directed against AchR, the natural next question is what causes these antibodies to appear? In short, we do not know, but there are some observations that may be regarded as clues. One such relates to immunogenetics. Some 20%–30% of MG patients have HLA–A1, B8, and/or DRw3. The majority of these are female, less than 40 years of age, with intermediate levels of AchR-specific antibodies, antibodies to muscle striations, and intermediate frequency of other autoantibodies. Another 30%–40% of MG patients have HLA-A3, B7, and/or DRw2 and are mostly elderly males with lower levels of AchR-specific antibodies and a high frequency of other autoantibodies (COMPSTON et al. 1981). Furthermore a DQ_β-associated restriction fragment length polymorphism has led to the identification of a 15-kb restriction fragment which, in DR3-positive individuals, is associated with a high risk for developing MG. These linkages between HLA phenotype and MG indicate that there may be a genetic predisposition to the development of MG and, furthermore, that genetic background may dictate the age of onset and the spectrum and intensity of the associated autoimmune phenomena.

Another relevant observation is that patients with MG have antibodies in their sera directed against the same epitopes as animals immunized with AchR (TZARTOS et al. 1981). This has been interpreted by some as suggesting that the immune response against AchR in MG patients may be initiated by an exposure to AchR rather than by an exposure to cross-reacting determinants on another molecule. There are, however, alternative interpretations, and one of them is that most of the AchR-specific antibodies in the sera of MG patients are secondary to destruction of the AchR and release of antigenic degradation products. In support of this possibility are the increased AchR degradation and decreased AchR content in striated muscles of patients with MG and the lack of correlation between the titer of AchR-specific antibodies and severity of MG. The first observation indicates that there may be more release of immunogenic components of AchR in patients with MG than in control populations. The second observation tells us that it is likely that most of the AchR-specific antibodies in MG have nothing to do with the pathogenesis of the disease and, therefore, could be secondary phenomena.

Thymoma occurs in 10%–20% of patients with MG, and these patients usually have high titers of AchR-specific antibodies and antibodies against muscle striations but no other autoantibodies (COMPSTON et al. 1981). The possibility has been raised that in these cases the thymoma may, through molecular mimicry, be responsible for initiating the pathogenic autoimmune response against the AchR. This assertion is entirely unproven. However, MG patients with thymoma have characteristic autoantibodies against striated muscle. AARLI et al. (1987) have reported that 85% of MG patients with thymoma have in their sera antibodies against a citric acid extract from striated muscle, whereas only 8% of MG patients without thymoma do. This indicates that the thymoma or events associated with or leading to thymoma do induce a distinct autoimmune response against striated muscle, and a part of this response may be in the form of AchR-specific antibodies that lead to MG. It is not clear how the autoimmune response against the striated muscle and the thymoma are linked, but one possibility is that the autoimmune response is elicited by epitopes on the thymoma that are shared by the striated muscle. If this is the case and MG in patients with thymoma is caused by molecules in the thymoma that mimic AchR, what causes MG in the 80%–90% of patients who do not have thymoma? Our hypothesis is that some (or all) of these cases of MG may be traced to an exposure to microorganisms carrying molecules that, with respect to immune surveillance, mimic the AchR. In the following pages, we will review the evidence that we and others have gathered in support of this hypothesis. We will discuss sharing of epitopes by the α-chain of the AchR and viruses, the sharing of epitopes by the α-chain of the AchR and gram-negative bacteria, and finally the way in which interaction of bacteria with the idiotypic network may lead to induction of autoantibodies against the AchR.

2 Sharing of Epitopes by Viruses and the Acetylcholine Receptor

OLDSTONE and his colleagues (DYRBERG and OLDSTONE 1986) approached the question of the pathogenesis of MG in the same manner they successfully used to approach the question of the pathogenesis of postinfectious encephalomyelitis (FUJINAMI and OLDSTONE 1985). They did a computer search in a protein sequence data base (Protein Identification Resource, Washington DC) for viral protein sequences similar to the α-chain of the AchR. They placed most emphasis on finding similarity to an eight amino acid stretch in the α-chain that some investigators believe constitutes the major immunogenic portion of the AchR (NODA et al. 1983). The philosophy behind this approach is that, although all attempts to demonstrate preceding or concurrent viral infections in MG patients have failed (and may continue to do so), it is possible that the autoimmune response against the AchR may be elicited by viral components that mimic the AchR. Conceivably, all traces of a preceding viral infection may be cleared before MG becomes clinically apparent, and once the tolerance of the immune system for the cross-reacting self-antigen (AchR) has been broken, the immune response against AchR may continue in the absence of the inciting virus.

Some of the sequences they selected for further study are listed in Table 1.

Peptides were synthesized with these sequences, coupled to keyhole limpet hemocyanin, and used to immunize rabbits. The antibodies raised in this manner were

Table 1. Comparison of sequences similar to the α-chain of acetylcholine receptor

Protein	Amino Acid Positions	Amino Acid Sequence
Human AchR α-chain	160–167	P E S D Q P D L
Polyomavirus middle T antigen	317–324	P E S D Q D Q L
Herpes simplex virus glycoprotein D	286–293 381–388	P N A T Q P E L R E D D Q P S S
Parvovirus H1.VP2 protein	135–142	T E T N Q P D T

then tested for reactivity with the peptide from human AchR. There was considerable variation in antibody cross-reactivity, which apparently correlated poorly with the degree of sequence similarity between the viral peptides and the AchR peptide. For example, two distinct eight residue peptides from herpes simplex virus glycoprotein D, each sharing four amino acid residues with the AchR peptide, were used as immunogens. Antibodies against one were found to be strongly reactive with the AchR peptide, but antibodies against the other did not cross-react. Another surprise was that antibodies generated against the peptide with the greatest similarity to the AchR peptide (polyomavirus middle T antigen) did not react with the AchR peptide.

Thus, there are viral proteins that contain stretches of sequence similar to what could be a major immunogenic region of the α-chain of the AchR (NODA et al. 1983). Furthermore, antibodies raised against some of these viral peptides react with the AchR peptide. However, we do not know whether the antibodies raised against the viral peptides react with the intact AchR, nor do we know whether the segments of the viral proteins that mimic the AchR are immunogenic. Finally, it is unclear whether the presence of antibodies elicited by the viral proteins and cross-reacting with residues 160–167 in the AchR would lead to MG. However, the mere existence of the shared epitopes by the α-chain of the AchR and certain viral proteins commands attention and gives a clear direction for future research. It is of interest here that, when viral proteins containing sequences similar to the AchR peptide are analyzed by computer programs that map hydrophilicity and predict flexibility in polypeptide chains (Intelligentics, CA), the viral sequences are all in fairly flexible and hydrophilic regions of the proteins. This indicates that the viral peptides may be immunogenic. In contrast, the segment 160–167 from the α-chain of the AchR is in an inflexible and hydrophobic region of the molecule. This could be viewed as a reason to anticipate that the viral sequences would elicit autoimmune responses, but AchR would not.

3 Sharing of Epitopes by the α-Chain of the Acetylcholine Receptor and Gram-Negative Enteric Bacteria

There are numerous examples of epitopes shared by bacteria and host tissues (SODERSTROM et al. 1984; FINNE et al. 1983), some of which may participate in the pathogenesis of autoimmune diseases. The classic example is provided by *Streptococcus pyogenes* and its post-infectious complications, rheumatic fever and glomerulonephritis. It has been shown that the M protein of *S. pyogenes* shares epitopes with cardiac myosin (KRISCHER and CUNNINGHAM 1985) and a component of renal glomerulae (KRAUS and BEACHEY 1988). Another recent example (ARCHER et al. 1985; SCHWIMMBECK et al. 1987) demonstrates the sharing of an epitope(s) by HLA B27 and the nitrogenase from *Klebsiella pneumoniae*. Antibodies against this shared epitope exist in the sera of patients with spondylitis ankylopoietica and patients with Reiter's syndrome.

In our search for evidence to support the hypothesis that molecular mimicry may be involved in the pathogenesis of autoimmune diseases, we analyzed micro-

Table 2. Approximate molecular weights of bacterial polypeptides having cross reactive determinants with acetylcholine receptor

Bacteria used	Polypeptides identified by BK57 (approx. molecular weight)	Polypeptides identified by 77F (approx. molecular weight)
E. coli	38000	—
	55000	—
K. pneumoniae	38000	—
	55000	—
P. vulgaris	55000	55000
Y. enterocolitica	38000	—
S. fexneri	—	—
P. aeruginosa	—	—
C. perfringens	—	—
S. aureus	—	—
S. pyogenes	—	—
P. putridas	—	—
B. subtilis	—	—

organisms for molecules sharing epitopes with AchR. Our approach was to test on Western blots a library of monoclonal antibodies (mAbs) against AchR for reactivity with electrophoresed polypeptides from bacteria. Since MG has not been tied to any specific infectious illness, we believed it was plausible to begin the search among bacteria that belong to normal gut flora. We did, however, extend the search to pathogenic bacteria.

Two of 50 mAbs reacted with polypeptides in the bacteria we tested (see Table 2); both are IgMs and are directed against the α-subunit of AchR. The mAb BK57 reacted with two polypeptides from *Eschericia coli* and *Klebsiella pneumoniae*, one with a molecular weight of 38000 and the other, of 55000. The mAb BK57 also reacted with one polypeptide from *Yersinia enterocolitica* and one from *Proteus vulgaris* (Fig. 1). The mAb BK57 reacted with nothing in *Shigella flexneri*, *Streptococcus*

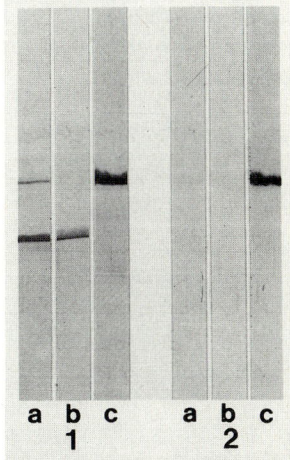

Fig. 1. Western blot of polypeptides from *E. coli* (*a*), *Y. enterocolitica* (*b*), and *P. vulgaris* (*c*). Group *1* was incubated with mAb BK57, group *2* with mAb 77F. Binding was visualized by indirect peroxidase. BK57 binds to polypeptides from *E. coli* and *Y. enterocolitica* having an approximate molecular weight (MW) of 38000 and to polypeptides in *E. coli* and *P. vulgaris* having an approximate MW of 55000. 77F binds only to a polypeptide in *P. vulgaris* having a MW of approximately 55000

pyogenes, Staphylococcus aureus, Pseudomonas aeruginosa, Clostridium perfringens, Bacillus subtilis, or *Peptostreptococcus putridus*. The other mAb, 77F, reacted with a polypeptide in *P. vulgaris* but reacted with no other bacteria tested, including *E. coli*. Hence, we showed that two polypeptides in *E. coli* and one polypeptide in *Y. enterocolitica* share at least one epitope with the α-chain of the AchR and that one polypeptide in *P. vulgaris* shares at least two epitopes with the α-chain. The polypeptides in *E. coli* that react with mAb BK57 are both membrane proteins (STEFANSSON et al. 1985). The smaller one is in the outer membrane and is the same size as a wellcharacterized outer membrane protein called ompC (LUGTENBERG 1981).

Sera from a few patients with MG were shown to contain antibodies that reacted strongly on immunoblots with two polypeptides from *Y. enterocolitica* (STEFANSSON et al. 1987). One of these polypeptides was isolated by preparative electrophoresis, and the isolated protein was tested for reactivity with antibodies from the sera of MG patients. Sera from MG patients were shown to contain antibodies against this polypeptide in higher titers than controls. We want to emphasize that it was not shown that the antibodies that bind to the polypeptide from *Y. enterocolitica* similarly bind to AchR. Furthermore, we do not know whether the epitopes shared by the bacterial polypeptides and the α-chain of the AchR are immunogenic or if antibodies against them cause disease.

4 Antigens on Bacteria Could Cause Myasthenia Gravis Through Sharing of Epitopes by Antibodies Against Bacterial Components and the Acetylcholine Receptor

The variable region of an antibody that contains the antigen-binding site, the paratope, can also be viewed as an antigen or a collection of epitopes. These epitopes comprise the idiotype of the antibody. Epitopes within an idiotype are referred to as idiotopes (JERNE 1974). In his theory about the idiotypic network, Jerne postulated that antibodies directed against some idiotopes participate in the regulation of humoral immune responses. There is now considerable experimental evidence in support of this notion (EICHMANN 1978; BONA 1981; RAJEWSKY and TAKEMORI 1983). To understand what follows, it is important to keep in mind that antibodies with different specificities (paratopes) may share idiotopes. Such idiotopes may behave as epitopes, which in turn may be present on the targets of autoimmune attack.

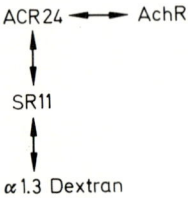

Fig. 2. The mAb *SR11* binds to α1.3 dextran, and *ACR24* binds to the variable portion of SR11. ACR24 also binds to the acetylcholine receptor (*AchR*). This shows clearly that humoral immune responses against the two antigens could be linked

DWYER et al. (1986) have shown in an elegant study that AchR and α1.3 dextran are linked through an idiotypic network. They raised mAbs against α1.3 dextran and against AchR as well as mAbs specific to the variable regions of the first mAbs (anti-ids). A simplified version of this work follows (Fig. 2): An mAb raised against the AchR, ACR24, reacts with the variable portion of an mAb, SR11, which was raised against α1.3 dextran. ACR24 is an anti-id for SR11.

The α1.3 dextran is present on certain bacteria including *Enterobacter cloacae* and *Serratia liquefaciens*. This observation led to the hypothesis that an exposure to bacteria displaying α1.3 dextran could lead to α1.3 dextran-specific antibodies, which in turn would lead to anti-id antibodies. These anti-id antibodies could then react with AchR and cause MG. In further support of this hypothesis, the authors showed that 12 of 60 MG patients had antibodies against α1.3 dextran in their sera, whereas none of the 40 controls did. However, the authors do not claim that the dextran-specific antibodies in the sera of MG patients induce antibodies to AchR nor have they shown that immunization with α1.3 dextran leads to experimental MG in animals.

5 Conclusions

We have briefly reviewed three lines of work that clearly show how an exposure to certain viruses or bacteria could lead to the formation of antibodies against AchR either through epitopes shared by the microorganisms and AchR or through epitopes shared by antibodies against the microorganisms and AchR. This invites the hypothesis that at least a certain proportion of MG may be caused by an exposure to such microorganisms. It is likely that this exposure would have to occur on the background of a genetic predisposition. As mentioned in the introduction, it is an appealing possibility that MG in patients with thymoma may be caused by molecules on the thymoma that share epitopes with the AchR. Still another example of how molecular mimicry may enter into hypotheses about the pathogenesis of MG is found in the work of LEFVERT et al. (1987), who have reported several cases of MG in recipients of bone marrow transplants. It is tempting to postulate that the autoimmune response against AchR is, in this instance, caused by epitopes on polymorphic antigens in the marrow (HLA antigens, minor antigens) that are shared by AchR.

In summary, there is substantial circumstantial evidence in support of the hypothesis that molecular mimicry may lead to MG, even possibly in the majority of patients. However, it has yet to be proven that it ever does.

References

Aarli JA, Gilhus NE, Hofstad H (1987) CA-antibody: an immunological marker of thymic neoplasia in myasthenia gravis? Acta Neurol Scand 76 (1): 55–57

Archer JR, Stubbs MM, Currey HLF, Gecry AF (1985) Antiserum to *Klebsiella* K43 BTS1 specifically lyses lymphocytes of HLA-B27 positive patients with ankylosing spondylitis from a London population. Lancet 1: 344–345

Bell JI, Steinman L, Toyka K, McDevitt HO (1987) HLA-DQ restriction fragment length polymorphisms in myasthenia gravis. Ann NY Acad Sci 505: 382–387

Bona CA (1981) Immune network: regulation of lymphocyte functions by anti-idiotypic antibodies. In: Idiotypes and lymphocytes. Academic, New York, pp 156–182

Compston DAS, Vincent A, Newsome-Davis J, Batchelor J (1981) Clinical, pathological, HLA antigen and immunological evidence for disease heterogeneity in myasthenia gravis. Brain 103: 579–601

Dwyer DS, Vakil M, Kearney JF (1986) Idiotypic network connectivity and a possible cause of myasthenia gravis. J Exp Med 164: 1310–1318

Dyrberg T, Oldstone MBA (1986) Peptides as probes to study molecular mimicry and virus-induced autoimmunity. Curr Topics Microbiol Immunol 130: 25–37

Eichmann K (1978) Expression and function of idiotypes on lymphocytes. Adv Immunol 26: 195–254

Finne J, Leinonen M, Makela PH (1983) Antigenic similarities between brain components and bacteria causing meningitis: implication for vaccine development and pathogenesis. Lancet 2: 355–357

Fujinami RS, Oldstone MBA (1985) Amino acid homology between the encephalitogenic site of myelin basic protein and virus: mechanism for autoimmunity. Science 230: 1043–1045

Gomez CM, Richman DP, Burres SA, Arnason BGW (1981) Monoclonal hybridoma anti-acetylcholine receptor antibodies: antibody specificity and effect of passive transfer. Ann NY Acad Sci 377: 97–109

Grob D, Brunner NG, Namba T (1981) The natural course of myasthenia gravis and effect of therapeutic measures. Ann NY Acad Sci 377: 652–669

Herrmann D Jr (1967) The first three centuries. Myasthenia Gravis Bull, Los Angeles Neurol Soc 32: 131

Jerne NK (1974) Towards a network theory of the immune system. Ann Immunol (Paris) 125C: 373–389

Kraus W, Beachey EH (1988) Renal autoimmune epitope of group A streptococci specified by M protein tetrapeptide Ile-Arg-Leu-Arg. Proc Natl Acad Sci USA 85: 4516–4520

Krisher K, Cunningham MW (1985) Myosin: a link between streptococci and heart. Science 227: 413–415

Lefvert AK, Bolme P, Hammerstrom L, Lonnqvist B, Ringden O, Slordahl S, Smith CIE (1987) Bone marrow grafting selectively induces the production of acetylcholine receptor antibodies, immunoglobulins bearing related idiotypes, and antiidiotypic antibodies. Ann NY Acad Sci 505: 825–827

Lindstrom JM, Engel AG, Seybold ME, Lennon VA, Lambert EH (1976a) Pathological mechanisms in EAMG. II. Passive transfer of experimental autoimmune myasthenia gravis in rats with anti-acetylcholine receptor antibodies. J Exp Med 144: 739–753

Lindstrom JM, Seybold ME, Lennon VA, Whittingham S, Duane D (1976b) Antibody to acetylcholine receptor in myasthenia gravis: prevalence, clinical correlates and diagnostic value. Neurology 26: 1054–1059

Lindstrom JM (1979) Autoimmune response to acetylcholine receptor in myasthenia gravis and its animal model. Adv Immunol 27: 1–50

Lugtenberg B (1981) Composition and function of the outer membrane of *Escherichia coli*. TIBS 6: 262–266

Noda M, Furutani Y, Takahashi H, Toyosato M, Tanabe T, Shimizu S, Kikyotani S, Kayano T, Hirose T, Inayama S, Numa S (1983) Cloning and sequence analysis of calf cDNA and human genomic DNA encoding alpha-subunit precursor of muscle acetylcholine receptor. Nature 305: 818–823

Patrick J, Lindstrom JM (1973) Autoimmune response to acetylcholine receptors. Science 180: 871–872

Rajewsky K, Takemori T (1983) Genetics, expression and function of idiotypes. Ann Rev Immunol 1: 569–607

Schwimmbeck PL, Yu DTY, Oldstone MBA (1987) Autoantibodies to HLA B27 in the sera of HLA B27 patients with ankylosing spondylitis and Reiter's syndrome. J Exp Med 166: 173–181

Soderstrom T, Hansson G, Larson G (1984) The *Escherichia coli* K1 capsule shares antigenic determinants with the gangliosides GM3 and GD3. N Engl J Med 310: 726–727

Stefansson K, Dieperink ME, Richman DP, Gomez CM, Marton LS (1985) Sharing of antigenic determinants between the nicotinic acetylcholine receptor and proteins in *Escherichia coli*, *Proteus vulgaris*, and *Klebsiella pneumoniae*. N Engl J Med 312 (4): 221–225

Stefansson K, Dieperink ME, Richman DP, Marton LS (1987) Sharing of epitopes by bacteria and the nicotinic acetylcholine receptor: a possible role in the pathogenesis of myasthenia gravis. Ann NY Acad Sci 505: 451–460

Tzartos SJ, Seybold M, Lindstrom JM (1981) Specificity of antibodies to acetylcholine receptors in the sera from myasthenia gravis patients measured by monoclonal antibodies. Proc Natl Acad Sci USA 79: 188–192

Celiac Disease: Adenovirus and Alpha Gliadin

M. F. KAGNOFF

1 Introduction   67
2 Etiology and Pathogenesis   68
3 Dietary Grains That Activate Celiac Disease   68
4 Possible Role of an Intestinal Adenovirus Protein   70
4.1 Sequence Similarity Between A-gliadin and the Adenovirus 12 E1b Protein   70
4.2 Evidence for Prior Exposure to Adenovirus 12   73

References   77

1 Introduction

Celiac disease (gluten-sensitive enteropathy, celiac sprue, nontropical sprue) is characterized by damage to the small intestinal mucosa and malabsorption of most nutrients (reviewed in COLE and KAGNOFF 1985; KAGNOFF 1989). The disease is activated by dietary exposure to wheat gluten and similar proteins in several other grains. Wheat gluten is a mixture of gliadin and glutenin (KASARDA 1981). It is the gliadin fraction that is responsible for activating this disease (KASARDA 1981). The symptoms commonly appear during the first 3 years of life after the introduction of cereals into the diet, with a second peak incidence occurring during the 3rd and 4th decades. Clinical manifestations predominantly reflect the consequences of malabsorption. Treatment consists of a gluten-free diet (i.e., a diet free of gluten and related, disease-associated grains) (Cole and Kagnoff 1985).

The pathogenesis of celiac disease appears to involve interactions between environmental, genetic, and immunologic factors (Kagnoff 1988, 1989). In terms of environmental factors, it is well-recognized that the disease can be activated by defined proteins present in several dietary grains, including wheat, rye, barley, and oats (reviewed in COLE and KAGNOFF 1985). The major genetic association of celiac disease has been with gene products of the HLA class II D region of the major histocompatibility locus complex on chromosome 6 (ALPER et al. 1987; HOWELL et al. 1986, 1988). Humoral and cell-mediated immune responses have been the focus of investigation in many studies that have attempted to define an immunologic basis for this disease (reviewed in COLE and KAGNOFF 1985; KAGNOFF 1988, 1989). This

---

Laboratory of Mucosal Immunology, Department of Medicine, University of California at San Diego, La Jolla, CA 92093, USA

chapter addresses the possibility that environmental factors other than dietary grains and, in particular, a protein encoded by a human adenovirus that is usually isolated from the intestinal tract may contribute to the pathogenesis of celiac disease.

2 Etiology and Pathogenesis

Several competing hypotheses have been put forth to explain the etiology and pathogenesis of celiac disease. The possibility that it was caused by a primary deficiency of a small intestinal peptidase, with incomplete hydrolysis of gluten and direct gluten toxicity to the mucosa, was popular for several years. However, this does not appear to be the case (STERCHI and WOODLEY 1978). Thus, the small intestinal mucosa contains multiple peptide hydrolases with overlapping substrate specificities, and biochemical abnormalities in intestinal peptidase activity in individuals with active celiac disease usually return to normal when they are treated with a gluten-free diet. Based on the observation that gliadin treated with a carbohydrase enzyme did not activate disease in several patients, carbohydrate side chains on the gliadins were postulated to be important in disease pathogenesis (PHELAN et al. 1977). However, this does not appear to be the cause of celiac disease, since several major α-gliadin components known to activate the disease lack carbohydrate side chains (BERNARDIN et al. 1976). In addition, little to no evidence supports a more recent hypothesis that purified gluten has lectin-like properties that are important in disease pathogenesis (KOLBERG and SOLLID 1985). Evidence also argues against the hypothesis that a primary defect in intestinal mucosal permeability is responsible for celiac disease (BJARNASON and PETERS 1984). For example, abnormal mucosal permeability may normalize on a gluten-free diet (HAMILTON et al. 1982). As discussed in the remaining portion of this chapter, the most compelling current models of disease pathogenesis view celiac disease as an immunologic disease in which environmental, genetic, and immunologic factors each contribute to the disease process (KAGNOFF 1988, 1989).

3 Dietary Grains That Activate Celiac Disease

Celiac disease is activated when a genetically susceptible host ingests food products that contain wheat, rye, barley, or oats. It is the gliadin fraction of wheat gluten and similar alcohol-soluble proteins (termed prolamins) in the other grains that are associated with the development of intestinal damage.

Cereal grains belong to the grass family (Gramineae). As shown in Fig. 1, grains other than wheat that activate celiac disease (e.g., rye and barley) have a close taxonomic relationship to wheat. Oats, which in large quantities are thought to activate this disease, are further removed from wheat, rye, and barley. Grains that do not activate the disease (e.g., rice and corn) are still further separated from wheat in their

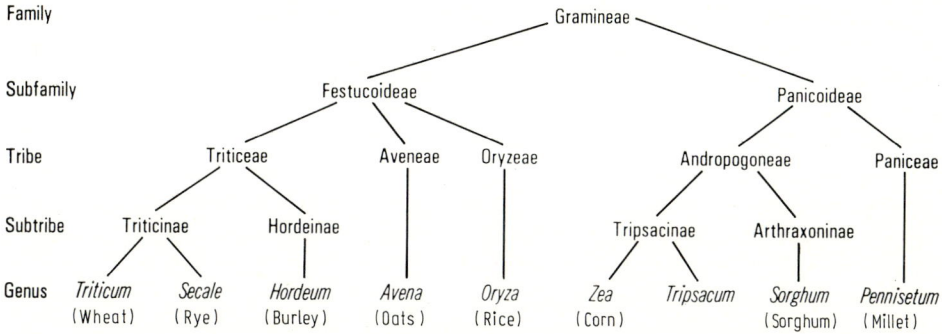

Fig. 1. Taxonomic relationships of major cereal grains. (Adapted from BIETZ, 1982)

derivation from the primitive grasses Common bread wheats developed relatively recently in evolution (i.e., the past 10000 years). They have a hexaploid genome (i.e., genome content AABBCC). Durum wheat, commonly used in the production of pasta, is tetraploid (i.e., genome content AABB), whereas rye and barley are diploid (KASARDA et al. 1976b; KASARDA 1981). Gene clusters that code for the gliadins in wheat are present on chromosomes of homologous groups 1 and 6. Because wheat gliadins are coded for on more than one chromosome, the breeding of wheat varieties that lack disease-activating properties has not been practical (CICLITIRA et al. 1980).

Gluten is a major component of the wheat endosperm and provides a source of nitrogen for the germinating wheat embryo (KASARDA et al. 1976a; KASARDA 1981). Gliadins and glutenins (mostly low molecular weight glutenins) are the major protein components of gluten, but only the gliadins appear to activate celiac disease. Minor constituents present in gluten extracts (e.g., wheat albumins, globulins, membrane proteins, lipids, and carbohydrates) also do not appear important in disease activation. The elastic properties of gluten are important in the production of bread dough.

Gliadins are single polypeptide chains that range in molecular weight from 30000 to 75000. They have a low charge and a very high glutamine and proline content (32–56 glutamine and 15–30 proline residues per 100 amino acid residues) (KASARDA et al. 1976a). Gliadin from a single variety of wheat may contain 40 or more different, but closely related, components (KASARDA et al. 1976a) when examined by two-dimensional gel electrophoresis. Gliadins, by convention, have been categorized into four major electrophoretic fractions, based on their migration properties in aluminum lactate gels at pH 3.2: alpha (α), beta (β), gamma (γ), and omega (ω) (KASARDA et al. 1976a). Each fraction in turn contains several subcomponents (e.g., β_1, β_2, β_3-gliadins; γ_1, γ_2, γ_3-gliadins; ω_1, ω_2, ω_3, ω_4-gliadins). Gliadins of the α, β, and γ_1 fraction share a similar amino acid composition and N-terminal sequence (i.e., α-type sequence) (KASARDA 1981). γ_2, γ_3, and ω-gliadins differ markedly from the α-type sequence in their N-terminal sequence and amino acid composition (i.e., γ-type sequence).

A-gliadin is a major α-gliadin component that is known to activate celiac disease (FALCHUCK et al. 1974; HEKKENS et al. 1970; KASARDA 1981). Recently, the complete primary amino acid sequence of A-gliadin was determined from amino acid sequencing (KASARDA et al. 1984), and other α-gliadin and γ-gliadin sequences have been deduced from sequencing of cDNA clones (BARTELS and THOMPSON 1983; RAFALSKI et al. 1984). The issue of which wheat gliadin fractions can activate disease is not settled. Early reports based on in vitro tests of damage to intestinal biopsies in organ culture and on in vivo tests in patients suggested that only α-gliadins or β, and perhaps, but not ω-gliadins might activate disease (KENDALL et al. 1972; JOS et al. 1978). Because ω-gliadins have the highest content of glutamine and proline, a high content of these amino acids alone is not thought to be the determining factor in disease activation. However, recent but controversial reports based on in vivo challenge studies in a few patients and in vitro organ culture studies suggested that all gluten fractions might activate disease (CICLITIRA et al. 1984; HOWDLE et al. 1984; JOS et al. 1982). Complete hydrolysis of gliadin destroys its disease-activating properties.

Controversy over which gliadin fractions activate celiac disease may stem in part from (a) significant heterogeneity in the sensitivity among patients to different gliadin fractions, (b) differences in the doses and purity of gliadin preparations used in challenge studies, (c) differences in the time after in vivo gliadin challenge at which small intestinal mucosal biopsies for diagnosis are obtained, (d) the use of different clinical and diagnostic endpoints to assess mucosal damage after in vivo gliadin challenge, or (e) an erroneous assumption that abnormalities in in vitro assays can be equated with disease activation.

4 Possible Role of an Intestinal Adenovirus Protein

There is a lack of complete concordance for celiac disease among monozygotic twin pairs. Thus, at least twenty-four pairs of "identical" twins with celiac disease have been reported (POLANCO et al. 1981), of which 18 (75%) were concordant for disease. Not all had monozygosity unequivocally proven, and some twin pairs have not had sufficient long-term follow-up to be certain that the disease will not develop at a later age. Nonetheless, it appears that there are cases of discordance for celiac disease among monozygotic twins. Based on these findings and the assumption that twins ingest similar dietary grains, we suggested that environmental factors other than dietary grains may be important in the pathogenesis of celiac disease (KAGNOFF 1984).

4.1 Sequence Similarity Between A-gliadin and the Adenovirus 12 E1b Protein

To explore the above possibility, we initially examined known protein sequences in the databank at the University of California, San Diego, for amino acid sequence homology with A-gliadin, a major α-gliadin component known to activate disease. A-gliadin has a molecular mass of 31 000 daltons, and its complete primary amino acid sequence was determined from amino acid sequencing and the sequencing of a cDNA clone (KASARDA et al. 1984). Our search to date has involved more than 4000 protein sequences. As initially noted, A-gliadin has a region of amino acid sequence homology with the E1b protein of human adenovirus serotype 12 (Ad12) (KAGNOFF 1984).

The region of homology between A-gliadin and the 54K Ad12 A1b protein spans 12 amino acids and includes 8 amino acid residue identities and an identical pentapeptide (Fig. 2). A-gliadin has 32 glutamines and 15 prolines per 100 amino

Fig. 2. Amino acid sequence of the adenovirus 12 E1b protein (*Ad12, E1b*) and A-gliadin beginning at amino acid residues 384 and 206, respectively. The region of homology includes 8 of 12 residue identities, including an identical pentapeptide. The single-letter code for amino acids is used. (From KAGNOFF 1984)

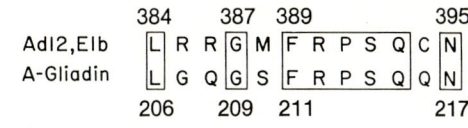

acid residues (KASARDA et al. 1984). However, the region of sequence similarity with the E1b protein involves domain V of A-gliadin, a domain that lacks repeating sequences with a high glutamine and proline content (KASARDA et al. 1984). In addition, the region includes only a single glutamine and proline residue in the Ad12 E1b protein. The amino acid similarity between regions of these proteins is probably due to chance, because A-gliadin and the Ad12 E1b protein are unrelated functionally and are not likely to share a common ancestry.

As shown in Fig. 3, the region of sequence homology between A-gliadin and the Ad12 E1b protein is hydrophilic in both proteins, suggesting that those sequences may be located on the exterior of the respective proteins (KYTE and DOOLITTLE 1982). Further studies have indicated that antisera raised against the 54K Ad12 E1b protein

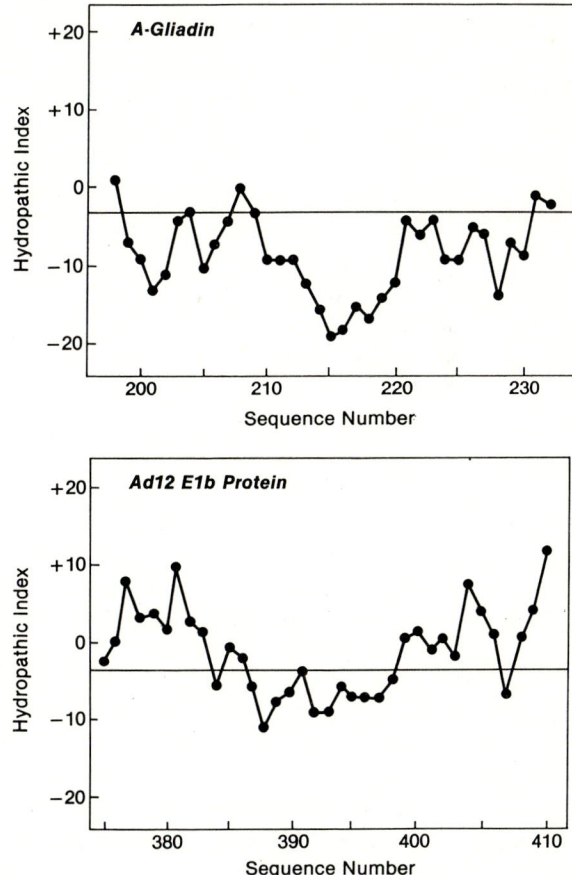

Fig. 3. Hydropathic character of A-gliadin over the segment spanning residues 198–232 (*top*) and of adenovirus 12 E1b protein over the segment spanning residues 375–410 (*bottom*). (From KAGNOFF 1984)

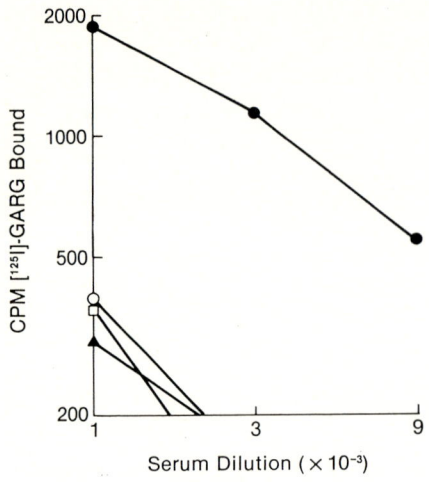

Fig. 4. Antiserum to adenovirus 12 E1b protein reacts with a synthetic A-gliadin peptide spanning residues 211–217. For the radioimmunoassay, wells of microtiter plates were sensitized with the A-gliadin heptapeptide coupled to bovine serum albumin. Reaction with anti Ad12 E1b sera (●); two different anti Ad5 E1b sera (○, □), or a control rat serum (▲)

specifically cross-reacts with A-gliadin and, as shown in Fig. 4, a heptapeptide of A-gliadin (FRSPQQN) spanning residues 211–217. Although the E1b protein from Ad5 is highly homologous with the Ad12 E1b protein (Bos et al. 1981; KIMURA et al. 1981), it does not share a region of sequence similarity with A-gliadin, and antisera to Ad5 E1b do not cross-react with A-gliadin or the synthetic heptapeptide (Fig. 4). Taken together, these studies indicate that antibody raised to the native 54K Ad12 E1b protein can specifically cross-react with A-gliadin in the region of shared sequence. Conversely, affinity-purified rabbit antibody to A-gliadin was shown to cross-react with Ad12 E1b protein prepared from an Ad12-transformed cell line (KAGNOFF 1984).

Ad12 is a double-stranded DNA virus that is usually isolated from the human intestinal tract and has been detected in stool samples as early as the first 1–2 years of life (FLINT 1980; MIDDLETON 1982). Ad12 has not been implicated previously as a cause of human disease but has been studied extensively because of its ability transform mammalian cell lines (FLINT 1980). Thus, cells transformed by Ad12 induce tumors in rodents at a high frequency within a relatively short time period (MAK et al. 1979). The transforming activity of Ad12 has been assigned to the left-hand 11% of the viral genome (i.e., early region I) (FLINT 1980), which contains two transcirptional units, E1a and E1b in a 3.9-kb DNA segment (PERRICAUDET et al. 1980; WILSON et al. 1979). Those units are the first to be expressed during lytic infection of human cells by Ad12. Of note, the 54K protein is not a structural protein of Ad12. However, it does represent the predominant virus-encoded protein expressed in the cytoplasm of mammalian cell lines that have been transformed by Ad12 (FLINT 1980).

4.2 Evidence for Prior Exposure to Adenovirus 12

Based on the studies described above, we asked whether patients with celiac disease had evidence of prior infection with Ad12. For these studies, Ad12-neutralizing antibody was used as an indication of past exposure to Ad12. Ad12-neutralizing antibody is directed to determinants on the structural hexon (ε) protein of Ad12 (NORRBY and ANKERST 1969), a protein that is not related to the Ad12 E1b protein or A-gliadin. As shown in Fig. 5, 89% (16/18) of a group of untreated celiac disease

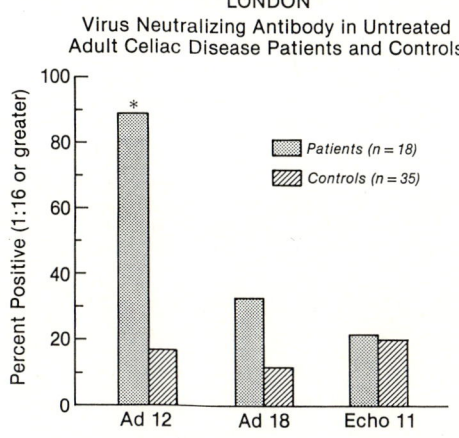

Fig. 5. Adenovirus and echovirus neutralizing antibody among untreated adult celiac disease (CD) subjects and controls in London. *n*, number of subjects in each group; *asterisk* indicates a value significantly different from control group ($P < 0.05$). (From KAGNOFF 1987)

patients from London had neutralizing antibody to Ad12 (KAGNOFF 1987). In contrast, such antibody was present in only 17% (6/35) of disease controls from the same institution (Fig. 5). Antibody to Ad12 in the celiac disease subjects did not appear to reflect a general increase in antibody titers to multiple viruses in those individuals. Thus, there was no significant difference in the prevalence of neutralizing antibody to echovirus 11 among individuals with celiac disease, compared with disease controls (Fig. 5). In addition, there was no significant difference in the presence of neutralizing antibody to Ad18, an adenovirus closely related to Ad12, between subjects with untreated celiac disease and controls. Although antibody to Ad18 was present in a greater number of untreated adult celiac disease subjects in London compared with controls, those data did not achieve statistical significance. The prevalence for Ad18-neutralizing antibody in control populations previously reported (D'AMBROSIO et al. 1982) approximated that noted among several control groups we have studied, ranging from 5% to 20%.

Ad12-neutralizing antibody titers for individual subjects are shown in Table 1. Of 6 subjects with celiac disease and positive titers to Ad18, titers to Ad18 exceeded those to Ad12 in two subjects, whereas in the remaining four subjects, titers to Ad12 exceeded those to Ad18.

Among patients treated for celiac disease in San Diego County, 33.3% (10/30) had positive Ad12-neutralizing antibody titers, in contrast to none of a group of matched healthy controls ($P < 0.05$). In other studies, we found neutralizing antibody to Ad12 in 30.8% (4/13) of a group of children treated for celiac disease in London, compared with 12.8% (9/70) of disease controls ($P < 0.05$). The prevalence of antibody to Ad12 in our control adults and children, including those with intestinal disorders other than celiac disease, approximated the 0%–15% prevalence noted before by others for Ad12-neutralizing antibody among different control populations (D'AMBROSIO et al. 1982; KAGNOFF 1987).

Further studies in patients with untreated celiac disease have demonstrated serum antibodies to a 6-amino acid synthetic peptide (RPSQQN) of A-gliadin from within

Table 1. Reciprocal neutralizing antibody titers to adenoviruses 12 and 18 in untreated and treated celiac disease subjects and controls[a]

Untreated			Treated			Controls		
Subject No.	Ad12	Ad18	Subject No.	Ad12	Ad18	Subject No.	Ad12	Ad18
1	128	16	1	16	<2	1	32	32
2	128	4	2	8	<2	2	32	16
3	64	<2	3	4	<2	3	32	<2
4	64	<2	4	4	<2	4	16	16
5	64	<2	5	2	16	5	16	16
6	32	4	6	2	2	6	16	8
7	32	8	7	2	2	7	8	16
8	32	16	8	2	<2	8	8	16
9	32	16	9	2	<2	9	8	8
10	32	16	10	2	<2	10	4	4
11	32	4	11	2	<2	11	4	2
12	32	4	12	<2	16	12	4	4
13	16	2	13	<2	2	13	4	<2
14	16	2	14	<2	2	14	4	4
15	16	<2	15	<2	<2	15	2	2
16	16	<2	16	<2	<2	16	2	2
17	8	16	17	<2	<2	17	2	2
18	4	16	18	<2	<2	18	2	2
			19	<2	<2	19	2	2
						20	2	2
						21	2	<2
						22	2	<2
						23	2	<2
						24	2	<2
						25	2	<2
						26	<2	<2
						27	<2	2
						28	<2	2
						29	<2	<2
						30	<2	<2
						31	<2	<2
						32	<2	<2
						33	<2	<2
						34	<2	<2
						35	<2	<2

[a] Titers greater than 8 were regarded as positive.

the region of sequence shared by A-gliadin and the Ad12 Elb protein, although such antibodies appear to be of low affinity (Fig. 6). Individuals lacked detectable antibody to irrelevant control synthetic peptides of similar size. In contrast to patients with active disease, patients treated for celiac disease have not had detectable serum antibody to peptides from the region shared by A-gliadin and the Ad12 Elb protein (KAGNOFF 1987).

These observations suggest that Ad12 may play a role in the pathogenesis of celiac disease, possibly due to chance immunologic cross-reactivity between shared epitopes on an Ad12-encoded protein and components of the α-gliadin fraction of wheat gluten.

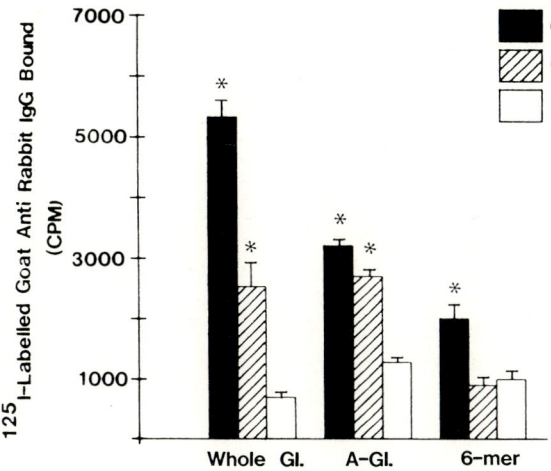

Fig. 6. Solid phase radioimmunoassay for antibody to *whole gliadin*, *A-gliadin*, and a synthetic 6-amino acid peptide, spanning residues 212–217 of A-gliadin in untreated and treated celiac disease subjects and controls. *n*, number of subjects in each group. Values are means, and *bars* represent ± SEM. *Asterisks* indicate values significantly different from control group ($P < 0.05$)

In support of this notion, most patients with active celiac disease in the group examined had evidence of past Ad12 exposure. Further, neutralizing antibody titers to Ad12 were increased significantly among adults and children with treated celiac disease compared with healthy or disease controls. The greater prevalence of neutralizing antibody to Ad12 in the untreated than treated patients is of interest and may reflect more recent infection with Ad12, although our studies have not directly addressed this issue. We also note that these studies do not exclude the possibility that patients with celiac disease, perhaps by virtue of an as yet undescribed abnormality in the intestinal mucosa, may be more likely than controls to acquire Ad12 infection.

In summary, these studies establish that A-gliadin has a region of amino acid sequence similarity with the E1b protein of Ad12. Further, they document an increased prevalence of past infection with Ad12 among patients with celiac disease and demonstrate that the region of sequence similarity between A-gliadin and the Ad12 E1b protein can act as a determinant for antibody recognition. Others have demonstrated that synthetic A-gliadin peptides from this region also can act as antigenic determinants in in vitro assays of cell-mediated immunity (KARAGIANNIS et al 1987). Thus, a 12-amino acid synthetic peptide (i.e., A-gliadin residues 206–217) from this region was reported to inhibit leukocyte migration in an indirect assay of leukocyte migration in patients treated for celiac disease, and to stimulate the release of a lymphokine with the properties of leukocyte migration inhibition factor in patients treated for celiac disease (LYDFORD-DENIS et al. 1987). Parenthetically, we note that patients with dermatitis herpetiformis (DH) also have a significantly increased prevalence of Ad12 neutralizing antibody (M. Kagnoff and I. Gigli, personal communication). This latter finding is of particular interest, since most patients with DH have a celiac disease-like intestinal lesion that responds to gluten withdrawal from the diet, and, like celiac disease, there is a marked association of DH with the HLA antigens -DR3 and -DQw2 (KATZ et al. 1980; LAWLEY et al. 1980).

We have hypothesized that shared epitopes between a protein coded for by a virus and determinants on α-gliadins may play a role in the pathogenesis of celiac disease by virtue of molecular mimicry, most likely at the level of T-cell recognition.

A striking feature of celiac disease is its association with a specific HLA haplotype, characterized serologically by the HLA class II antigens -DR3, -DR7, and -DQw2 (ALPER et al. 1987; TOSI et al. 1983) and a 4-kb *Rsa*I DPβ-chain genomic restriction fragment length polymorphism (HOWELL et al. 1988). Peptide fragments associated with specific HLA class II molecules on antigen-presenting cells form a bimolecular complex that is recognized by the receptor for antigen on CD4 T cells (SCHWARTZ 1985). The strong association between celiac disease and specific HLA class II molecules, taken together with the present data, suggest that peptides shared by A-gliadin and a virally encoded protein can cross-react at the level of T-cell recognition.

One consequence of such cross-priming of T cells could be the generation of CD4 T cells that help in the induction of antibody responses to antigenic determinants in the region of shared sequence, as we have detected, as well as to determinants on other parts of the α-gliadin molecule. Further, if the region of homology between A-gliadin and the Ad12 E1b protein acts as a T helper cell determinant, antibodies to epitopes on α-gliadins that cross-react with γ- and ω-gliadins and barley or rye prolamins, as is commonly seen in patients with this disease (LEVENSON et al. 1985), would be predicted. Such antibodies could play a role in disease pathogenesis by one of several mechanisms. For example, gliadin-specific antibody complexed with gliadin to form immune complexes could activate tissue-damaging effector mechanisms, including the complement cascade. Alternatively, gliadin-specific antibody could cause intestinal injury via a cell-mediated cytotoxic reaction in which gliadin-specific antibody recognizes gliadin peptides bound to mucosal structures and directs a killer (K) cell-mediated, antibody-dependent, cytotoxic reaction. In support of the above possibilities, IgG-producing cells capable of mediating such reactions are markedly increased in the lamina propria of the small intestinal mucosa during active disease (SCOTT et al. 1984), and organ culture studies have demonstrated increased local production of gliadin-specific antibody in celiac small intestinal mucosa after gliadin challenge (CICLITIRA et al. 1986). Nonetheless, current evidence neither documents nor refutes a major role for gliadin-specific antibody in the pathogenesis of disease. Alternatively, the peptide sequence shared by the E1b protein and A-gliadin could act as part of a target structure that is recognized on cells of the intestinal mucosa by cytotoxic cells of the immune system. Further testing of these possibilities is currently in progress.

Acknowledgements. We thank Ms. D. Sagall and Ms S. Lai for preparation of the manuscript. Work supported by NIH grant AM35108.

References

Alper CA, Fleischnick E, Awdeh Z, Katz AJ, Yunis EJ (1987) Extended major histocompatibility complex haplotypes in patients with gluten-sensitive enteropathy. U Clin Invest 79: 251–256

Bartels D, Thompson RD (1983) The characterization of cDNA clones coding for wheat storage proteins. Nucleic Acids Res 11: 2961–2977

Bernardin JE, Aunders RM, Kasarda DD (1976) Absence of carbohydrate in celaic-toxic A-gliadin. Cereal Chem 53: 612–614

Bietz JA (1982) Cereal prolamin evolution and homology revealed by sequence analysis. Biochem Genet 20: 1039–1053

Bjarnason I, Peters TJ (1984) In vitro determination of small intestinal permetability: demonstration of a persistent defect in patients with celiac disease. Gut 25: 145–150

Bos JL, Polder LJ, Bernards R, Schrier PI, van den Elsen PJ, van der Eb AJ, van Ormondt H (1981) The 2.2 kb E1b mRNA of human Ad12 and Ad5 codes for two tumor antigens starting at different AUG triplets. Cell 27: 121–131

Ciclitira PJ, Hunter JO, Lennox ES (1980) Clinical testing of bread made from nullisomac 6A wheats in celiac patients. Lancet ii: 234–236

Ciclitira PJ, Evans DJ, Fagg NLK, Lennox ES, Dowling RH (1984) Clinical testing of gliadin fractions in celiac patients. Clin Sci 66: 357–364

Ciclitira PJ, Ellis HJ, Wood GM, Howdle PD, Losowski MS (1986) Secretion of gliadin antibody by coeliac jejunal mucosal biopsies cultured in vitro. Clin Exp Immunol 64: 119–124

Cole SG, Kagnoff MF (1985) Celiac disease. In: Olson RE (ed) Annual Review of Nutrition. Annual Reviews Inc., Palo Alto, CA, pp 241–266

D'Ambrosio E, Del Grosso N, Chicca A, Midulla M (1982) Neutralizing antibodies against 33 human adenoviruses in normal children in Rome. J Hyg Camb 89: 155–166

Falchuk ZM, Gebhard RL, Sessoms C, Strober W (1974) An in vitro model of gluten sensitive enteropathy. Effect of gliadin on intestinal epithelial cells of patients with gluten-sensitive enteropathy in organ culture. J Clin Invest 53: 487–500

Flint SJ (1980) Structure and genomic organization of adenoviruses. In: Tooze J (ed) Molecular biology of tumor viruses II. DNA tumor viruses. Cold Spring Harbor Laboratory, New York, pp 383–441, 547–576

Hamilton I, Cobden I, Rothwell J, Axon ATR (1982) Intestinal permeability in celiac disease: the response to gluten withdrawal and single-dose gluten challenge. Gut 23: 202–210

Hekkens WTJM, Haex AJC, Willighagen RGJ (1970) Some aspects of gliadin fractionation and testing by a histochemical method. In: Booth CC, Dowling RH (eds) Coeliac disease. Churchill Livingstone, Edinburgh, pp 11–19

Howdle PD, Ciclitira PJ, Simpson FG, Losowsky MS (1984) Are all gliadins toxic in celiac disease? An in vitro study of alpha, beta, gamma, and omega gliadins. Scand J Gastro 19: 41–47

Howell MD, Austin RK, Kelleher D, Nepom GT, Kagnoff F (1986) An HLA-D region restriction fragment length polymorphism associated with celiac disease. J Exp Med 164: 333–338

Howell MD, Smith JR, Austin RK, Kelleher D, Nepom GT, Volk B, Kagnoff MF (1988) An extended HLA-D region haplotype associated with celiac disease. Proc Natl Acad Sci USA 85: 222–226

Jos J, Charbonnier L, Mosse J, Olives JP, de Tand M, Rey J (1982) The toxic fraction of gliadin digests in celiac disease. Isolation of chromatography on Biogel P-10. Clinica Chimica Acta 119: 263–274

Jos J, Charbonnier L, Mougenot JF, Mosse J, Rey J (1978) Isolation and characterization of the toxic fraction of wheat gliadin in celiac disease. In: McNicholl B, McCarthy CF, Fottrell PF (eds) Perspective in celiac disease. University Park Press, Baltimore, pp 75–90

Kagnoff MF (1984) Possible role for a human adenovirus in the pathogenesis of celiac disease. J Exp Med 160: 1544–1557

Kagnoff MF (1987) Evidence for the role of a human intestinal adenovirus in the pathogenesis of celiac disease. Gut 28: 995–1001

Kagnoff MF (1989) Celiac disease: pathogenesis and clinical features. In: Thomson ABR, Shaffer EA (ed) Modern concepts in gastroenterology. Plenum, New York (in press)

Kagnoff MF (1988) Celiac disease: a model of an immunologically-mediated disease. In: Kagnoff MF (ed) Immunology and allergy clinics of North America. WB Saunders, Philadelphia, 8: pp 505–520

Karagiannis JA, Priddle JD, Jewell DP (1987) Cell-mediated immunity to a synthetic gliadin peptide resembling a sequence from adenovirus 12. Lancet ii: 884–886

Kasarda DD (1981) Toxic proteins and peptides in celiac disease: relations to cereal genetics. In: Walcher D, Kretchmer M (eds) Food nutrition and evolution. Masson, New York, pp 201–216

Kasarda DD, Bernardin JE, Nimmo CC (1976a) Wheat proteins. In: Advances in cereal science and technology. American Association of Cereal Chemists, St Paul, MN, pp 158–236

Kasarda DD, Bernardin JE, Qualset CO (1976b) Relationship of gliadin protein components to chromosomes in hexploid wheats. Proc Natl Acad Sci USA 73: 3646–3650

Kasarda DD, Okita TW, Bernardin JE, Baecker PA, Nimmo CC, Lew EJ, Dietler MD, Greene FC (1984) Nucleic acid (cDNA) and amino acid sequences of alpha-type gliadins from wheat (*Triticum aestivum* L.). Proc Natl Acad Sci USA 81: 4712–4716

Katz SI, Hall RP, Lawley TJ, Strober W (1980) Dermatitis herpetiformis: the skin and the gut. Ann Int Med 93: 857–874

Kendall MJ, Cox PS, Schneider R, Hawkins CF (1972) Gluten subfractions in coeliac disease. Lancet 2: 1065–1067

Kimura T, Sawada Y, Shinawawa M, Shimizu Y, Shiroki K, Sugisaki H, Takanami M, Uemizu Y, Fujinaga K (1981) Nucleotide sequence of the transforming early region Elb of adenovirus type 12 DNA: structure and gene organization, and comparison with those of adenovirus type 5 DNA. Nucleic Acids Res 9: 6571–6589

Kolberg J, Sollid L (1985) Lectin activity of gluten identified as wheat germ agglutinin. Biochem Biophys Res Commun 130: 867–872

Kyte J, Doolittle RF (1982) A simple method for displaying the hydropathic character of a protein. J Mol Biol 157: 105–132

Lawley TJ, Strober W, Yaoita H, Katz, SI (1980) Small intestinal biopsies and HLA types in dermatitis herpetiformis patients with granular and linear IgA skin deposits. J Invest Dermatol 74: 9–12

Levenson SD, Austin RK, Dietler MD, Kasarda DD, Kagnoff MF (1985) Specificity of anti-gliadin antibody in celiac disease. Gastroenterology 89: 1–5

Lewis AM, Cook JL (1984) The interface between adenovirus-transformed cells and cellular immune response in the challenged host. Curr Topics Microbiol Immunol, 110: pp 1–22

Lydford-Davis H, Karagiannis JA, Priddle JD, Jewell DP (1987) Preliminary characterization of leukocyte migration inhibition factor (LIF) produced by lymphocytes from celiac patients when stimulated with gluten peptides. Clin Sci 72: 89P

Mak S, Mak I, Smiley JR, Graham FL (1979) Tumorigenicity and viral gene expression in rat cells transformed by Ad12 virions or by the EcoRI C fragment of Ad12 DNA. Virology 98: 456–460

Middleton PJ (1982) Role of viruses in pediatric gastrointestinal disease and epidemiologic factors. In: Tyrell DAJ, Kapikian AZ (eds) Virus infections of the gastrointestinal tract. Dekker, New York, pp 211–225

Norrby E, Ankerst J (1969) Biological characterization of structural components of adenovirus type 12. J Gen Virol 5: 183–194

Perricaudet M, le Moullec J-M, Tiollais P (1980) Structure of two adenovirus type 12 transforming polypeptides and their evolutionary implications. Nature 288: 174–176

Phelan JJ, Stevens FM, McNicholl B, Fottrell PF, McCarthy CF (1977) Celiac disease: the abolition of gliadin toxicity by enzymes from *Aspergillus niger*. Clin Sci Molec Med 53: 35–43

Polanco I, Biemond I, van Leeuwen A, Schreuder I, Meera, Kahn P (1981) Gluten-sensitive enteropathy in Spain: Genetic and environmental factors. In: McConnell RB (ed) The genetics of celiac disease. MTP Press, Lancaster, pp 211–231

Rafalski JA, Scheets K, Metzler M, Peterson DM, Hedgcoth C (1984) Developmentally regulated plant genes: the nucleotide sequence of a wheat gliadin genomic clone. EMBO J 3: 1409–1415

Schwartz RH (1985) T-lymphocyte recognition of antigen in association with gene products of the major histocompatibility complex. Annu Rev Immunol 3: 237–262

Scott BB, Goodall A, Stephenson P, Jenkins D (1984) Small intestinal plasma cells in celiac disease. Gut 25: 41–46

Sterchi EE, Woodley JF (1978) Peptidases of the human intestinal brush border membrane. In: McNicholl B, McCarthy CF, Fottrell PF (eds) Perspectivesin celiac disease. University Park Press, Baltimore, pp 437–449

Tosi R, Vismara D, Tanigaki N, Ferrara GB, Cicimarra F, Buffolano W, Follo D, Auricchio S (1983) Evidence that celiac disease is primarily associated with a DC locus allelic specificity. Clin Immunol Immunopathol 28: 395–404

Wilson MC, Fraser NW, Darnell JE (1979) Initiation sites by high doses of UV irradiation: evidence for three independent promoters within the left 11% of the Ad-2 gene. Virology 94: 175–184

Autoimmunity and Chagas' Disease

G. B. Takle[a,2] and L. Hudson[1,3]

1   Introduction   79
2   Pathology   80
2.1 Acute Chagas' Disease   80
2.2 Chronic Chagas' Disease   81
2.3 Acute Phase to Chronic Phase Transition   82
3   Mechanisms of Autoimmunity   82
3.1 Chronic Chagas' Disease and Autoimmunity   83
3.2 *Trypanosoma cruzi* Antigens   83
3.3 Laminin Cross-Reactivity   86
4   Vaccination and Autoimmunity   87
5   Expression of Parasite Antigens on Infected Cells   88
6   Conclusions   88
References   89

1 Introduction

The protozoan flagellate *Trypanosoma cruzi* is the causative agent of Chagas' disease and has been estimated to infect between 10 and 12 million people in Central and South America (WHO 1960). *T. cruzi* has a complex life cycle involving stages in both a vertebrate and an insect vector host, the reduviid or assassin bugs, members of the subfamily Triatominae. Transmission of infective metacyclic trypomastigotes to the vertebrate host occurs following feeding and defecation by the bug and contamination of a wound site or penetration of nearby mucous membranes by the parasite. *T. cruzi* is thus placed in the class Stercoraria in contrast to the pathogenic African trypanosome species, which are Salivaria, whose transmission to the vertebrate occurs by inoculation via the insect's salivary glands.

---

[1] Department of Immunology, St. George's Hospital Medical School, Cranmer Terrace, London SW17 ORE, UK

Present addresses:
[2] Department of Molecular Biology, Wellcome Biotech Ltd., Langley Court, Beckenham, Kent BR3 3BS, UK
[3] Division of Cellular Sciences, Glaxo Group Research Ltd., Greenford Road, London UB6 OHE, UK

Once in the vertebrate host, *T. cruzi* metacyclic trypomastigotes enter local tissue cells or macrophages, transform into rounded amastigotes in the cytoplasm, divide, and are released as motile trypomastigotes that serve to disseminate the infection. This early stage of the disease is often characterized by a local skin tissue lesion or 'chagoma' at the site of parasite entry and heralds the onset of the acute phase. Usually the acute phase of Chagas' disease is asymptomatic in adults, despite a detectable blood parasitemia. Occasionally, however, acute Chagas' disease pathology occurs in adults, or more commonly in children, and is accompanied by fever, enlarged spleen and lymph nodes, and myocarditis. In humans, the acute phase usually lasts between 1 and 2 months during which inflammatory damage to heart tissue may occur, although this usually resolves at the end of patent parasitemia due to the development of a vigorous immune response. A latent period or indeterminate phase follows (ANDRADE 1983) and may last for the remainder of the patient's life, providing the immune response remains intact (BAROUSSE et al. 1980). However, a proportion of patients develop chronic Chagas' disease about 10 or more years after initial infection. The chronic disease is of two forms: cardiac or digestive, the type and severity of which vary according to geographical region and have been correlated with *T. cruzi* strain (MILES et al. 1981).

2 Pathology

2.1 Acute Chagas' Disease

Chagas' disease pathology has been subject of an extensive review (SANTOS-BUCH 1979) to which the reader is referred for a complete appraisal of the subject. Acute Chagas' disease, involving fever, muscle pain, vomiting, diarrhea, enlarged lymphoid tissue, and myocarditis, is rare in adults. The existence of autoimmune damage at this stage of the disease is a possibility, since autoantibodies giving the characteristic EVI immunofluorescence patterns on heart tissue (endocardium, vascular tissue, and interstitia; COSSIO et al. 1974a, b) have been detected (Fig. 1). Most pathology, however, is probably related to parasite-associated local tissue damage and inflammation (ANSELMI and MOLEIRO 1974). After almost 4 months, patent high blood parasitemia subsides as patients become seropositive, and the disease enters the indeterminate phase. Parasites are detectable during this extended latent period by using sensitive multiplicative diagnostic methods such as xenodiagnosis or hemoculture. The indeterminate phase may last for up to 30 years in what is thought to be about 10% of those infected patients who go on to develop chronic Chagas' syndrome and ends when either the cardiac or the digestive abnormalities that typify the chronic disease appear.

According to ANSELMI and MOLEIRO (1974) the acute phase of Chagas' disease is characterized by cardiac pathology involving monocyte infiltration, degenerative myocarditis, and cyst formation. KOBERLE (1974) suggests that the extent of neuronal destruction in the acute phase determines the severity of the subsequent chronic phase symptoms. Thus, although acute Chagas' disease can be considered in most cases to

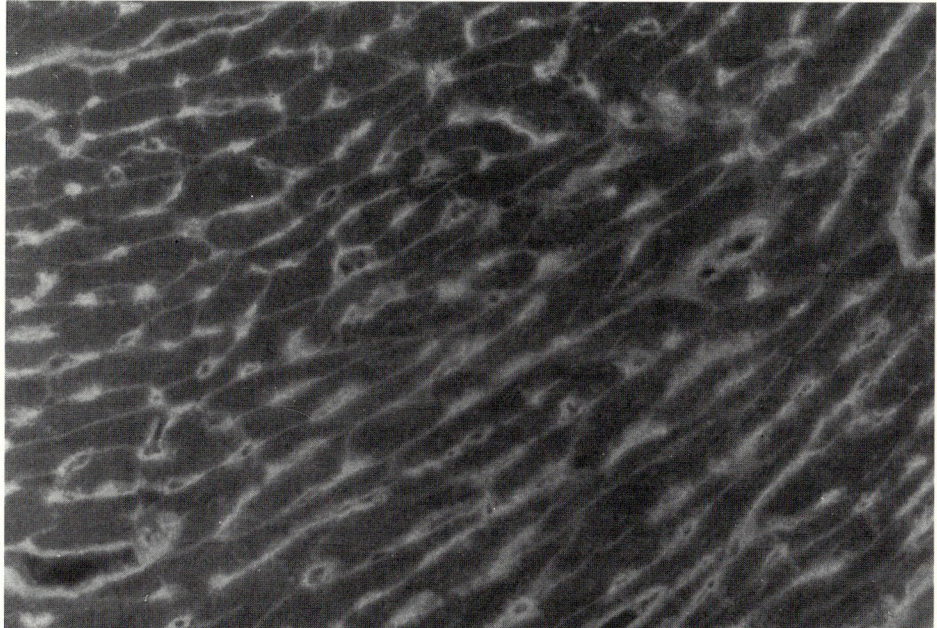

Fig. 1. EVI immunofluorescence. Serum from chronically infected chagasic patient used to stain cardiac tissue

be relatively benign, the damage that occurs early on in infection may have a profound and definitive influence on the outcome of the chronic disease.

2.2 Chronic Chagas' Disease

Destruction of neuronal tissue triggered by nearby lysing and necrosing infected muscle cells releasing parasites was observed as early as 1911 (VIANNA 1911) and is invariably followed by an array of intense inflammatory reactions. This damage of ganglion cells is certainly the cause of the various chronic Chagas' syndromes such as megacardia, megacolon, or megaesophagus that characterize the later phase of the disease, and denervation of the heart is by far the most widespread histopathological occurrence (KOBERLE 1974). The precise interdependence of parasite load and degree of pathogenesis in chronic Chagas' disease is still to be resolved unequivocally. MAEKELT (1970) showed that cardiac abnormalities in Chagas' disease could be correlated to the level of *T. cruzi*-specific antibodies. Since it is difficult to evaluate *T. cruzi* parasitemia directly, especially in the chronic phase, trypanosome-specific antibody titers are used as a measure of blood parasitemia. This assumption is, of course, now open to doubt in the face of evidence for the involvement of autoantibodies in the pathology of Chagas' disease. KOBERLE (1974) maintains that cardiac and digestive symptoms of Chagas' disease result from early neuropathy caused by high level parasitemia, resulting in denervation that becomes exacerbated with time as a result of the normal, age-related loss of neurons. The features characteristic of chronic Chagas' disease include reduced cardiac output, dilatation of the heart chambers, and left ventricular aneurysm (TEIXEIRA 1979; ANDRADE et al. 1978) and in the digestive form, disturbed persistalsis, hypertrophy, and dilatation of regions of the alimentary tract, particularly the colon and esophagus.

2.3 Acute Phase to Chronic Phase Transition

In general, the progression of Chagas' disease has formerly been treated as though it had two distinct stages (acute and chronic). Although it is possible to delimit a clear biphasic progression, the latent phase may only represent an asymptomatic period of transformation from the parasite-mediated acute form to the parasite-initiated, but immunologically based chronic disease (HUDSON and HINDMARSH 1985). Also, it is possible to detect minor cardiac alterations during the indeterminate phase presumptive of full-blown chronic chagasic cardiomyopathy, and this state is similar to that found in the chronic disease in the mouse (HUDSON and HINDMARSH 1985). Thus, it is likely that in this transformation from acute to chronic state lies the cue for the development of autoimmunity. Low parasitemia that may persist throughout the duration of the indeterminate phase, restricted to a low level by the vigorous immune response, might serve to perpetuate or exacerbate any cross-reactive autoimmune component. In this case, in patients who go on to show chronic Chagas' disease, the so-called latent period is probably not functionally latent, but rather is a period of gradually enhanced autoreactivity that culminates in either the cardiac or digestive manifestations characteristic of the chronic disease.

3 Mechanisms of Autoimmunity

Before discussing the information pertaining to chagasic autoimmunity, it is worthwhile briefly reviewing the current thinking on the origin of the autoimmune state. Autoimmunity can be thought of as the breakdown of self-tolerance and can be achieved in a number of ways. In the normal individual, populations of self-reacting lymphocytes and autoantigens are present and exposed to each other. A suppressor mechanism operates to prevent the onset of autoimmunity, and this involves different populations of T suppressor (T_s) cells. Any defect in the T_s activity would thus open the way for self-reacting lymphocytes to respond to autoantigen. There are numerous possible mechanism that can be postulated to explain the breakdown of tolerance, but the regulation of the activity of anti-self T and B lymphocytes, involving the clonal abortion of anti-self T helper cells, the involvement of T_s cells or T contrasuppressor cells, the action of idiotype networks or T cell bypass are all considered to be important aspects of the basic theme. Not only is autoantibody production likely to be enhanced during autoimmunity, but also autoreactive T cells are likely to be stimulated, leading to T-cell cytotoxicity. This is relevant to chronic Chagas' disease, since autoreactive cytotoxic T cells to heart myoblasts have been reported (LAGUENS et al 1988), and demyelination, often reported in Chagas' disease, has been shown to be T cell-mediated during allergic encephalomyelitis (BEN NUN et al. 1981).

Of course, the most direct route to autoimmunity involves an infecting microbe containing a host cross-reactive surface antigen so that microbe-specific antibody or microbe-specific cellular immunity is concurrently directed towards the pathogen and self. Alternatively, idiotype networks involving idiotype-specific antibodies, microbe-specific antibodies, and the various classes of regulatory T cells could be responsible for generating autoimmunity. Clearly, loss of control over any one of the many regulatory processes that constitute the tolerant state will ultimately result in autoimmune disease.

3.1 Chronic Chagas' Disease and Autoimmunity

The phenomenon of autoimmunity in Chagas' disease was first postulated with the discovery of autoantibodies to normal endocardium, vascular structures, and interstitia of cardiac muscle (EVI antibodies; COSSIO et al. 1974a). This functional class of antibodies in EVI serum is thought to act as a β adrenergic agonist (STERIN-BORDU et al. 1976). Numerous other reports have now described autoantibodies against Schwann cells (KHOURY et al. 1979), laminin (SZARFMAN et al. 1982; AVILA et al. 1984), striated muscle (COSSIO et al. 1974b), and neurons (RIBIERO DOS SANTOS et al. 1979) in Chagas' disease. Autoantibodies have been shown to define universal host/parasite epitopes by cross-adsorption, and, more definitively, cross-reactive monoclonal antibodies have been isolated (WOOD et al. 1982; SNARY et al. 1983).

The finding of common epitopes shared by host and parasite would allow several possible routes for the development of pathogenesis in chagasic autoimmunity. In its simplest form parasite-induced autoimmunity can result from a parasite antigen sharing an epitope or epitopes with a host component. Alternatively, host tissue damage might stimulate autoimmunity by exposing "altered" self to the host immune system, and along with this comes the possibility that host components are adsorbed onto the surface of the parasite (PEYROL et al. 1987; WILLIAMS et al. 1985; BRETANA and O'DALY 1976) or vice versa (RIBIERO DOS SANTOS and HUDSON 1980) and become antigenic in this form. There is evidence that these phenomena do occur in chronic Chagas' disease, and it is likely that all are contributing to the onset of the autoimmune state. Clearly an analysis of the antigenic makeup of the parasite might be of value at this point in an attempt to highlight potential candidates for cross-reactive autoantigens.

3.2 *Trypanosoma cruzi* Antigens

Unlike African trypanosomes, each trypomastigote clone of which possesses a uniform variant surface glycoprotein coat, each *T. cruzi* clone contains an array of different outer membrane antigenic polypeptides. Analysis of a number of *T. cruzi* surface antigens reveals that they are glycosylated (SNARY 1985) in common with most eukaryotic surface polypeptides. *T. cruzi* shows no evidence of antigenic variation of the kind displayed by African trypanosomes (SNARY 1980), since *T. cruzi* parasites are covered with a nonuniform antigenic surface, and the successive waves of parasitemia characteristic of individual *T. brucei* Variant Antigen Types (VATs) (BORST and CROSS 1982) do not occur. It would be of benefit to the parasite, however, if epitopes on the most immunogenic individual surface glycoproteins underwent some form of variation, but a definitive analysis of this still has to be carried out.

There are four major detectable glycoprotein surface antigens of *T. cruzi* (SNARY 1985) and a large number of less prominent ones, depending on the methods used to analyze them (SCHECHTER and NOGUEIRA 1988) and on the strain or zymodeme of parasite used. Glycoproteins of molecular weight 90K (SNARY and HUDSON 1979), 85K (KATZIN and COLLI 1983; OUAISSI et al. 1986a, b), 72K (SNARY et al. 1981), and 25K (SCHARFSTEIN et al. 1983) have been most thoroughly investigated, and the genes for glycoproteins of 90K (BEARD et al. 1988) and 85K (PETERSON et al. 1986) have been cloned and sequenced. At least one of these glycoproteins identified by one-dimensional SDS-PAGE can be resolved into a number of distinct components by two-dimensional electrophoresis (ANDREWS et al. 1984).

GP90 is present on all life cycle stages of *T. cruzi* and has been purified by lectin affinity chromatography (SNARY and HUDSON 1979). It protects mice from acute lethal infection (SNARY 1983), but protection is not sterile, and this would not augur well for its use as a vaccine in the case of this predominantly chronic autoimmune disease. The gene for GP90 has recently been cloned from an expression library screened with a monoclonal antibody (BEARD et al. 1988) and has been shown to be part of a tandemly repeated multigene family. In fact, most genes of *T. cruzi* so far examined show this type of genomic arrangement (REQUENA et al. 1988; MAIGNON et al. 1988). GP90 is unlikely to stimulate the onset of cross-reactive autoimmunity, since antibodies to GP90 do not cross-react with host tissues (SCOTT and SNARY 1979).

An insect stage (epimastigote)-specific glycoprotein of molecular weight 72K has been extensively studied after its initial characterization by cell surface labeling and immunoprecipitation (NOGUIERA et al. 1981; SNARY et al. 1981). It is involved in differentiation and might act as a lectin receptor in the insect midgut (SHER and SNARY 1982). A monoclonal antibody (WIC 29.26) binds to the carbohydrate side chains of this glycoprotein. The sugar side chain compositions have been determined (FERGUSON et al. 1983) and reveal that GP72 is unique since it contains rhamnose. These unusual features of the GP72 carbohydrate side chains probably confer high immunogenicity and may explain the high levels of GP72-specific antibody in human chagasic sera, even though the epimastigote stage is relatively rare in mammalian hosts. Again there is no evidence for cross-reactivity between self components and GP72 (NOGUIERA 1986), but since the immune response is clearly directed at the carbohydrate portion of the molecule, it may be that in *T. cruzi* carbohydrates comprise the most immunogenic antigens.

A group of glycoproteins of molecular weight approximately 85K by one-dimensional SDS-PAGE forms a prominent component of ^{125}I-surface labeled *T. cruzi* (BEARD et al. 1985; OUAISSI et al. 1986a, b; ANDREWS et al. 1984). This 85K group resolves into at least three spots by two-dimensional PAGE and represents proteins with pIs of 5.5, 5.0, and 6.3–7.5 (ANDREWS et al. 1984). Glycoproteins of molecular weight 85K have been shown to be involved in the penetration of parasites into monolayer culture cells (KATZIN and COLLI 1983; QUAISSI et al. 1986a, b; PETERSON et al. 1986), and OUAISSI and colleagues have shown that a GP85 is a fibronectin receptor (OUAISSI et al. 1986b). The sequencing of the gene for one of the GP85 molecules shows that the polypeptide contains a region of degenerate tandem repeats that may be responsible for the strong immunogenicity of the molecule. It is possible that these repeats are cross-reactive to human sequences, but oligonucleotides to the repeat region do not detect bands in Southern blots of human genomic DNA (Takle, unpublished data). Whether the GP85s currently being investigated are identical remains to be determined; they are from different strains of *T. cruzi* but appear functionally similar. Recently a 140K fibronectin receptor, integrin (TAMKUN et al. 1986), has been identified and cDNA clones isolated from chick fibrolasts. Again it would be of interest whether the *T. cruzi* fibronectin receptor is in any way homologous to this vertebrate receptor.

The possibility that host serum fibronectin bound to the *T. cruzi* fibronectin receptor might induce autoimmunity is worth considering here. Although host components certainly bind to the surface of the parasite (PEYROL et al. 1987; PRIOLI et al. 1988), it is not known whether this is a specific receptor-ligand interaction or the result of nonspecific adsorption. In the case of fibronectin, and also in that of a recently identified host serum *T. cruzi* neuraminidase-binding protein, cruzin (PRIOLI et al. 1988), the interaction is specific, but nonspecific adsorption is also known to occur (WILLIAMS et al. 1985; RIBIERO DOS SANTOS and HUDSON 1980). With respect to fibronectin, it is unlikely that fibronectin-specific autoimmunity is induced by serum fibronectin binding to parasites, since no fibronectin-specific antibodies have been detected in patients with Chagas' disease (SZARFMAN et al. 1982). In the case of cruzin, which despite its name is in fact human high density lipoprotein, anticruzin reactivity has yet to be found in chagasic serum.

T. cruzi has been shown to contain numerous heat-shock proteins (hsp) (DRAGON et al. 1987; ALCINA et al. 1988) and at least one (*T. cruzi* 85K hsp) is homologous with yeast and *Drosophila* hsp (DRAGON et al. 1987). Although, like one of the GP85s (PETERSON et al. 1986), this 85K polypeptide has been isolated from the Peru strain of *T. cruzi*, it is not trypomastigote specific and is not identical with the GP85 surface antigen. Since this *T. cruzi* hsp 85K gene was detected using human chagasic serum, it is feasible that there could be chagasic cross-reactivity with human hsp.

The use of recombinant DNA expression vectors is particularly relevant in the determination of immunogenic *T. cruzi* surface polypeptides using immune chagasic serum as probes. IBANEZ et al. (1987) carried out a comprehensive study of lambda gt11 clones recognized by infected human serum. The low rate of positive clones detected (only 53 positives from 3×10^5 recombinant clones) is in accord with the notion that the majority of antibodies in this serum were probably directed against carbohydrate epitopes. The isolated clones selected antibodies to surface antigens ranging in size from 205K to 85K. Selected antibodies were not tested for their cross-reactivity with host tissue. One surface antigen that is known to be cross-reactive with heart tissue is a 67K protein (SANTOS-BUCH 1979), but so far this molecule has not been isolated or chemically characterized. It is beyond the scope of this review to describe every *T. cruzi* surface antigen and its possible involvement in autoimmunity. Clearly there is variation in the range of parasite proteins between *T. cruzi* strains (WRIGHTSMAN et al. 1986) and zymodemes (MCDANIEL et al. 1986), and variation in the methods used to identify them (SCHECHTER and NOGUIERA 1988) certainly contributes to this. The different *T. cruzi* antigens are thus still incompletely characterized either with regard to their involvement in autoimmunity or as candidate vaccines.

3.3 Laminin Cross-Reactivity

The existence of EVI antibodies and antibodies to Schwann cells in Chagas' disease is well-established (Cossio et al. 1974a, b), and there is discussion about whether they can be correlated with the extent of chronic chagasic lesions (Szarfman et al. 1981). The basement membrane-directed antibody reactivity in EVI serum is towards the connective tissue glycoprotein, laminin, but not towards other components of connective tissue such as fibronectin, collagen types I, III, IV, and V,

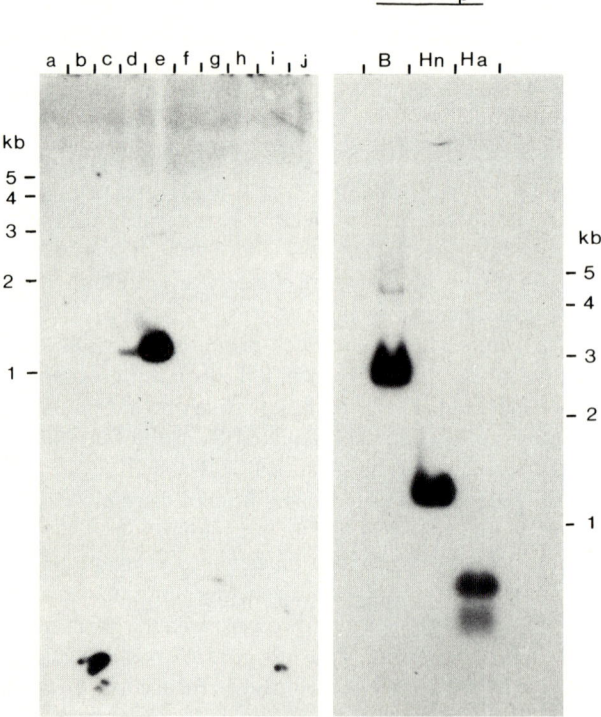

Fig. 2. *Left*, Southern blot using a 676–base pair mouse laminin B_2 subunit probe insert from clone pPE9 (Barlow et al. 1984). Genomic DNA (3 µg) was digested with *Hind*III (30 U) and run in a 0.8% agarose gel, blotted to Genescreen Plus, prehybridized, and hybridized according to manufacturers' instructions. Highest stringency washing used $0.1 \times SSC$ at 65 °C for 30 min. *a*, *T. cruzi* Y Paris trypomastigote; *b*, *T. cruzi* Y Paris epimastigote; *c*, *T. cruzi* Y clone epimastigote; *d*, *T. cruzi* California strain epimastigote; *e*, *T. cruzi* clone A21 epimastigote; *f*, human; *g*, *Toxoplasma gondii*; *h*, *Plasmodium yoelii*; *i*, *Babesia bovis*; *j*, *T. rangeli* San Augustin strain. *T. cruzi* clone A21 epimastigote DNA (3 µg) was digested with 30 U *Bam*H1 (*B*); *Hind*III (*Hn*); *Hae*III (*Ha*); and blotted, prehybridized, hybridized, and washed as above. Only a sequence in clone A21 is recognized by the mouse laminin probe and this sequence appears as a single copy

heparan sulfate, proteoglycan, or chondronectin (SZARFMAN et al. 1982). This finding was initially interpreted to imply that laminin-like molecules exist on the surfaces of *T. cruzi* trypomastigotes causing anti-laminin cross-reactivity. BRETANA et al. (1986) have identified such molecules by immunocytochemistry and have shown them to be trypomastigote specific and concentrated in particular regions of the plasma membrane. Culture forms of *T. cruzi* and the closely related South American trypanosome *T. rangeli* do not possess this group of molecules. The latter observation is surprising, since humans infected with the supposedly nonpathogenic *T. rangeli* alone have serum laminin-specific antibodies (AVILA et al. 1987). The implications of this finding are numerous. Firstly, since no laminin-like molecules are present on the surface of *T. rangeli* (BRETANA et al. 1986), the anti-laminin activity may be towards a host molecule. The most likely explanation is that chagasic autoantibodies to laminin are cross-reactive with carbohydrate epitopes, and evidence has been published to support this notion (TOWBIN et al. 1988). Laminin-specific autoantibodies react specifically with terminal Gal-α-1-3-Gal groups on N-linked oligosaccharides. Antibodies to this epitope are also found in normal human serum and also probably in *T. rangeli* infection serum, although the latter has not been investigated for Gal-α-1-3-Gal-specific activity. It may thus be that *T. rangeli* can be confused not only with a diagnosis of *T. cruzi* infection (GUHL et al. 1987) but could also stimulate a degree of autoimmunity and so might not be as harmless as previously thought.

Although laminin-specific antibodies in Chagas' disease are carbohydrate-specific, a mouse laminin B chain cDNA probe (BARLOW et al. 1984) does hybridize to genomic DNA from clone A21 of the Y strain of *T. cruzi* by Southern blotting, but not to other clones or strains (Fig. 2). This typifies the variable nature of the various *T. cruzi* strains and shows that the particular strain of parasite that brings about an infection could have profound consequences on the subsequent outcome or extent of autoimmunity.

4 Vaccination and Autoimmunity

Further analysis of possible cross-reacting molecules is important if we are to get a clearer understanding of the nature of chagasic autoimmunity. Potentially then, we are confronted with a "catch 22" situation when attempting to control Chagas' disease by vaccination since, in attempting to produce protective immunity using recombinant, synthetic peptide or other vaccines, the very response that serves to cure the infection may subsequently turn out to exacerbate autoimmunity. The need for careful analysis of the host response to all potential vaccines is thus self-evident.

Analysis by monoclonal antibodies provides a powerful method for investigating the properties of some immunogenic epitopes, and with this comes the possibility of using idiotypic-specific antibodies (Ab$_2$) to mimic these epitopes. WOOD et al. (1982) generated a monoclonal antibody (CE5) that is cross-reactive with mammalian neurons and *T. cruzi*. This monoclonal antibody shows a complex binding pattern in Western blots of parasite and neuronal tissue, indicating a common polypeptide or carbohydrate epitope. A complete determination of the nature of the epitopes would be of use to reveal possible universal autoimmune epitopes. Possibly this could be done by raising anti-idiotypes, inhibitory oligosaccharides or oligopeptides, or by using recombinant DNA methods.

In the case of a monoclonal antibody that recognizes a carbohydrate epitope on a major *T. cruzi* surface antigen (GP72), idiotype-specific antibodies have been raised and shown to induce high levels of Ab_2-specific reactivity (SACKS et al. 1985). This approach is useful in instances when vaccines based on carbohydrate epitopes could be important and again provides a technique for more definitive epitope mapping.

The finding that antibodies directed specifically against living trypomastigotes are associated with resistance to challenge by *T. cruzi* infection (KRETTLI and BRENER 1982; MARTINS et al. 1985) in humans and mouse has important implications for vaccine development. These protective antibodies specifically recognize an approximately 160K (glyco) protein (MARTINS et al. 1985), and we are currently undertaking a further characterization of this trypomastigote-specific 160K polypeptide. Obviously, there is a strong case for testing any 160K-specific antibodies against host tissue to check for cross-reactivity before its wide-scale use as a possible vaccine.

5 Expression of Parasite Antigens on Infected Cells

The phenomenon of cross-adsorption of parasite antigens and host tissue (RIBIERO DOS SANTOS and HUDSON 1980) has been described earlier. This phenomenon has hampered investigations aimed at characterizing the endogenous expression of parasite antigens by infected host cells. Certainly there is discussion about whether parasite-antigen expression on host cells does occur (BRENER 1980; HANSON 1977; ABRAHAMSON and KLOETZEL 1980), but ARAUJO (1985) provided clear evidence to support the notion of parasite (glyco) protein(s) expression on the host cell.

6 Conclusions

The existence of autoimmunity in chronic Chagas' disease is unequivocal, but the means by which chagasic autoimmunity develops and its importance in disease pathogenesis are still open to conjecture. There is now little doubt that some degree parasite/host cross-reactivity occurs, and definition of cross-reactive antigens and epitopes is now taking place (TOWBIN et al. 1987). It seems likely that a combination of events could result in cross-reactivity. The parasites themselves have cross-reactive surface molecules. Intracellular parasites lyse host cells; parasite material from this lysis adsorbs to surrounding host cells and activates self-directed immunity (RIBERO DOS SANTOS and HUDSON 1980). Infected cells express cross-reactive parasite antigens (ARAUJO 1985), and antigen-presenting cells can process and present cross-reactive parasite molecules. It is known that certain antigens are more effective cross-reactive targets for autoimmune states, because they are more easily degraded, and that the degradation products have higher affinity for MHC molecules (ADA and ROSE 1988). Possibly *T. cruzi* possesses a range of such antigens.

The observations that a relatively small and variable proportion of *T. cruzi*-infected patients progress to overt chronic chagasic megasyndromes suggest that there may be a genetic influence on the onset of autoimmunity. No thorough study of this nature has been carried out in humans. Yet, despite a lack of association between chronic Chagas' disease and HLA types (FERREIRA et al. 1979) in humans, in the mouse there is evidence for the involvement in acute Chagas' disease of host-restricted genetic elements (the *lpr* gene and Y_{SB}-associated factor; BOYER et al. 1983; NICKELL et al. 1985) that are known to influence the onset of autoimmunity. This is likely to be important in human Chagas' disease in which control over early histopathological events influences the extent of the chronic state.

Many facets of *T. cruzi* autoimmunity in chronic Chagas' disease still require investigation, and in this brief review we have attempted to highlight some areas in which progress has been made and also where knowledge is lacking. It seems to us that the mechanisms involved in chagasic autoimmunity are complex and numerous, requiring a rigorous experimental approach to rationalize each step and determine its clinical importance. Methods now available in immunochemistry and recombinant DNA technology not only facilitate this kind of approach but also allow optimism for a successful outcome.

Acknowledgements. We would like to thank the Wellcome Trust and the Medical Research Council for financial support, Dr. D. Snary for critically reading the manuscript, and Pam Sparks for typing. We are grateful to Dr. B. Hogan, NIMR, Mill Hill London for the gift of the laminin probe.

References

Abrahamson IA, Kloetzel JK (1980) Presence of *Trypanosoma cruzi* antigen on the surface of both infected and uninfected cells in tissue culture. Parasitology 80: 147–154
Ada GL, Rose NR (1988) The initiation and early development of autoimmune diseases. Clin Immunol Immunopathol 41: 3–9
Alcina, A., Urzainqui, A. and Carrasco, L. (1988). The heat shock response in *Trypanosoma cruzi*. Eur. J. Biochem. 172: 121–127
Andrade ZA (1983). Mechanisms of myocardial damage in *Trypanosoma cruzi* infection. Ciba Found Symp 99: 214–233
Andrade ZA, Andrade SG, Oliveira GB, Alonson DR (1978). Hisopathology of the conducting tissue of the heart in Chagas' myocarditis. Am Heart J 95 (3): 316–324
Andrews IW, Katzin AM & Colli W, (1984) Mapping of surface glycoproteins of *Trypanosoma cruzi* by two dimensional electrophoresis. Eur J Biochem 140: 599–604
Anselmi A and Moleiro F (1974) Pathogenic mechanisms in Chagas' cardiomyopathy. Ciba Found Symp 20: 125–136
Araujo FG (1985). *Trypanosoma cruzi*: expression of antigens on the membrane surface of parasitized cells. J Immunol 135: 4149–4154
Avila JL, Rojas M and Rieber M (1984). Antibodies to laminin in American cutaneous leishmaniasis. Infect Immun 43: 402–406
Avila JL, Rojas M, Velazquez-Avila G and Rieber M (1987). Antibodies to laminin in *Trypanosoma rangeli* infected subjects. Parasitol. Res. 73: 178–179
Barlow DP, Green NM, Kurkinen M and Hogan BLM (1984). Sequencing of laminin B chain cDNAs reveals C-terminal regions of coiled-coil alpha helix. EMBO J 3: 2355–2362
Barousse AP, Costa JA, Epasto, La Plume H and Segura EL (1980). Enfermedad de Chagas e immunosupresion. Medicina (Buenos Aires) 40: 17–26

Beard CA, Wrightsman RA and Manning JE (1985). Identification of monoclonal antibodies against the trypomastigote stage of *Trypanosoma cruzi* by use of iminobiotinylated surface polypeptides. Mol Biochem Parasitol 16: 199–212

Beard CA, Wrightsman RA and Manning JE (1988) Stage and strain specific expression of the tandemly repeated 90 kDa surface antigen gene family in *Trypanosoma cruzi*. Mol Biochem Parasitol 28: 227–234

Ben Nun A, Weckerle H and Cohen IR (1981) Vaccination against autoimmune encephalomyelitis with T-lymphocyte line reactive cells reactive against myelin basic protein. Nature 242: 60–61

Borst P and Cross GAM (1982) The molecular basis for trypanosome antigenic variation. Cell 29: 291–303

Boyer MH, Hoff R, Kipnis TL, Murphy ED & Roths JB (1983) *Trypanosoma cruzi*: susceptibility in mice carrying mutant gene lpr (lymphoproliferation). Parasite Immunol 5: 135–142

Brener Z (1980) Immunity to *Trypanosoma cruzi*. Adv Parasitol 18: 247–275

Bretana A and O'Daly JA (1976) Uptake of fetal proteins by *Trypanosoma cruzi* immunofluorescence and ultrastructural studies. Int Parasitol 6: 379–386

Bretana A, Avila JH, Arias-Flores M, Contreras M and Tapia FJ (1986) *Trypanosoma cruzi* and *Leishmania spp*: immunocytochemical localisation of a laminin-like protein in the plasma membrane. Exp Parasitol 61: 168–175

Cossio PM, Diez C, Szarfman A, Kreutzer E, Candiolo B and Arana RM (1974a) Chagasic cardiomyopathy: demonstration of a serum gamma globulin factor which reacts with endocardium and vascular structures. Circulation 49: 13025

Cossio PM, Laguens RP, Kreutzer E, Diez C, Segal A and Arana RM (1974b) Chagasic cardiopathy: antibodies reacting with plasma membrane of striated muscle and endothelial cells. Circulation 50: 1252–1263

Dragon EA, Roberts S, Kato EA and Gabe JD (1987) The genome of *Trypanosoma cruzi* contains a constitutively expressed tandemly arranged multicopy gene homologous to a major heat shock protein. Mol Cell Biol 7: 1271–1275

Ferguson MAJ, Allen AK and Snary D (1983) Studies on the structure of a phosphoglycoprotein from the parasitic protozoan *Trypanosoma cruzi*. Biochem J 207: 171–174

Ferreira E, Neva F, Gusmao R, Ward F, Rezende J, Rossi A, Amos DB and Johnson AH (1979) HLA and Chagas' disease. Abs Congresso Internacional Sobre Doenca DeChagas, Rio de Janeiro, Brasil, p 181

Guhl F, Hudson L, Marinkelle CJ, Jaramillo CA and Bridge D (1987) Clinical *Trypanosoma rangeli* infection as a complication of Chagas' disease. Parasitology 94: 475–484

Hanson WL (1977) Immune response and mechanisms of resistance in *Trypanosoma cruzi*. In: Chagas' disease. Pan American Health Organization, scientific publication no 343. PAHO, Washington DC p 22

Hudson L and Hindmarsh PJ (1985) The relationship between autoimmunity and Chagas' disease: causal or coincidental? Curr Top Microbiol Immunol 117: 167–177

Ibanez CF, Affranchino JL and Frasch ACC (1987) Antigenic determinants of *Trypanosoma cruzi* defined by cloning of parasite DNA. Mol Biochem Parasitol 25: 175–184

Katzin AM and Colli W (1983) Lectin receptors in *Trypanosoma cruzi*: an N-acetyl-D-glucosamine containing surface glycoprotein specific for the trypomastigote stage. Biochim Biophys Acta 727: 403–411

Khoury EL, Ritacco V, Cossio PM, Laguens RP, Szarfman A, Diez C and Arana RM (1979) Circulating antibodies to peripheral nerve in American trypanosomiasis (Chagas' disease). Clin Exp Immunol 36: 8–15

Koberle F (1974) Pathogenesis of Chagas' disease. Ciba Found Symp 20: 136–158

Krettli AV and Brener Z (1982) Resistance against *Trypanosoma cruzi* associated to anti-living trypomastigote antibodies. J Immunol 128: 2009–2012

Laguens RP, Meckert PC, Chambo JG (1988) Antiheart antibody-dependent cytotoxicity in the sera of mice chronically infected with *Trypanosoma cruzi*. Infect Immun 56: 993–997

Maeckelt GA (1970) Seroepidemiology of Chagas' disease. J Parasitol 56: 557–563

Maignon R, Gerke R, Rodriguez M, Urbina J, Hoenicka J, Neury S, Aguirre T, Nehlin J, Knapp T and Crampton J (1988) The tubulin genes of *Trypanosoma cruzi*. Eur J Biochem 171: 285–291

Martins MS, Hudson L, Krettli AV, Cancado JR and Brener Z (1985) Human and mouse sera recognise the same polypeptide associated with immunological resistance to *Trypanosoma cruzi* infection. Clin Exp Immunol 61/7: 343–350

McDaniel JP, Howard RJ and Dvorak JA (1986) Identification and analysis of epimastigote surface and metabolic proteins in *Trypanosoma cruzi*. Mol Biochem Parasitol 19: 183–194

Miles MA, Cedillos RA, Povoa MM, de Souza AA, Prata A and Macedo V (1981) Do radically dissimilar *Trypanosoma cruzi* strains (zymodemes) cause Venezuelan and Brazilian forms of Chagas' disease? Lancet i: 1338

Nickell SP, Hoff R and Boyer MH (1985) Susceptibility to acute *Trypanosoma cruzi* infection in autoimmune strains of mice. Parasite Immunol 7: 377–386

Nogueira N (1986) American Trypanosomiasis: Antigens and Host-Parasite Interactions. In, Receptors and Ligands in Intercellular Communication, Vol. 7. Parasite Antigens: Towards New Strategies for Vaccines. T. W. Pearson (ed) 4 pp 91–110. Marcel Dekker Inc. New York. Basel

Nogueira N, Chaplan S, Tydings JD, Unkless J and Cohn Z (1981) *Trypanosoma cruzi* surface antigens of blood and culture forms. J Exp Med 153: 629–639

Ouaissi MA, Cornette J and Capron A (1986a) Identification and isolation of *Trypanosoma cruzi* trypomastigote cell surface protein with properties expected of a fibronectin receptor. Mol Biochem Parasitol 19: 201–211

Ouaissi MA, Cornette J, Afchain D, Capron A, Gras-Masse H and Tartar A (1986b) *Trypanosoma cruzi* infection inhibited by peptides modelled from a fibronectin cell attachment domain. Science 234: 603–607

Peterson DS, Wrightsman RA and Manning JE (1986) Cloning of a major surface antigen gene of *Trypanosoma cruzi* and identification of a nona peptide repeat. Nature 322: 566–568

Peyrol S, Ouaissi MA, Capron A and Grimaud JA (1987) *Trypanosoma cruzi*: ultrastructural visualisation of fibronectin bound to culture forms. Exp Parasitol 63: 112–144

Prioli RP, Rosenberg I, Shivakumar S, Pereira MEA (1988) Specific binding of human plasma high density lipoprotein (cruzin) to *Trypanosoma cruzi*. Mol Biochem Parasitol 28: 257–264

Requena JM, Lopez MC, Himenez-Ruiz A, de la Torre JC and Alonso C (1988) A head-to-tail tandem organisation of hsp 70 genes in *Trypanosoma cruzi*. Nucleic Acids Res 16: 1393–1406

Ribiero dos Santos R and Hudson L (1980) Immunological consequences of parasite modification of host cells. Clin Exp Immunol 40: 36–41

Ribiero dos Santos R, Marquez JO, Von Gal Furtado CC, Ramos de Oliveira JC, Martins AR and Koberle F (1979) Antibodies against neurons in chronic Chagas' disease. Tropenmed Parasitol 30: 19–23

Sacks DL, Kirchhoff LV, Hieny S and Sher A (1985) Molecular mimicry of a carbohydrate epitope on a major surface glycoprotein of *Trypanosoma cruzi* by using anti-idiotype antibodies. J Immunol 135/7: 4155–4159

Santos Buch CA (1979) American trypanosomiasis: Chagas' disease. Int Rev Exp Pathol 19: 63–95

Scharfstein J, Rodrigues MM, Alves CA, De Souza W, Previato JO and Mendonca-Previato L (1983) *Trypanosoma cruzi*: description of a highly purified surface antigen defined by human antibodies. J Immunol 131: 972–976

Schechter M and Nogueira N (1988) Variations induced by different methodologies in *Trypanosoma cruzi* surface antigen profiles. Mol Biochem Parasitol 29: 37–46

Scott MT and Snary D (1979) Protective immunisation of mice using cell surface glycoprotein from *Trypanosoma cruzi*. Nature 282: 73–74

Sher A and Snary D (1982) Specific inhibition of the morphogenesis of *Trypanosoma cruzi* by a monoclonal antibody. Nature 300: 639–640

Snary D (1980) *Trypanosoma cruzi*: antigenic invariance of the cell surface glycoprotein. Exp Parasitol 49: 68–77

Snary D (1983) Cell surface glycoproteins of *Trypanosoma cruzi*: protective immunity in mice and antibody levels in human chagasic sera. Trans R Soc Med Hyg 77: 126–129

Snary D (1985) The cell surface of *Trypanosoma cruzi*. Curr Top in Microbiol Immunol 117: 75–92

Snary D and Hudson L (1979) *Trypanosoma cruzi* cell surface proteins: identification of one major glycoprotein. FEBS Lett 100: 166–170

Snary D, Ferguson MAJ, Scott MT and Allen AK (1981) Cell surface antigens of *Trypanosoma cruzi*:

use of monoclonal antibodies to identify and isolate an epimastigote specific glycoprotein. Mol Biochem Parasitol 3: 343–356

Snary D, Flint JE, Wood NJ, Scott MT, Chapman MD, Dodd J, Jessell TM and Miles MA (1983) A monoclonal antibody with specificity for *Trypanosoma cruzi*: central and peripheral neurons and glia. Clin Exp Immunol 54: 617–624

Sterin-Bordu L, Cossio PM, Gimeno MF, Gimeno AL, Diez C, Laguens RP, Meckert PC, Arana RM (1976) Effect of chagasic sera on the rat isolated atrial preparation: immunological, morphological and functional aspects. Cardiovasc Res 10: 613–622

Szarfman A, Luguettia A, Rossi A, Rezende JM and Schmunis GA (1981) Tissue reacting immunoglobulins in patients with different clinical forms of Chagas' disease. Am J Trop Med Hyg 30: 43

Szarfman A, Terranova VP, Rennard SI, Foidart JM, Lima MDF, Scheinman JI and Martin GR (1982) Antibodies to laminin in Chagas' disease. J Exp Med 155: 1161–1171

Tamkun JW, De Simone DW, Fonda D, Patel RS, Buck C, Horwitz AF and Hynes RO (1986) Structure of integrin, a glycoprotein involved in the transmembrane linkage between fibronectin and actin. Cell 46: 271–282

Teixeira ARL (1979) Chagas' disease: trends in immunological research and prospects for immunoprophylaxis. Bull WHO 57: 697–714

Towbin H, Rosenfielder G, Wieslander J, Avila JL, Rojas M, Szarfman A, Esser K, Nowack H and Timpl R (1987) Circulating antibodies to mouse laminin in Chagas' disease, American cutaneous leishmaniasis and normal individuals recognise terminal galactosyl α(1-3)-galactose epitopes. J Exp Med 166: 419–432

Vianna G (1911) Contribuicoes para o estudo de anatomia patalogiacada molestia de Carlos Chagas. Mem Inst Oswaldo Cruz 3: 276–294

WHO (1960) Chagas' disease report of a study group. WHO tech rep ser no 20

Williams GT, Fielder L, Smith H and Hudson L (1985) Adsorption of *Trypanosoma cruzi* proteins to mammalian cells in vitro. Acta Tropica 42: 33–38

Wood JN, Hudson L, Jessell TM and Yamamoto M (1982) A monoclonal antibody defining determinants on sub populations of mammalian neurones and *Trypanosoma cruzi* parasites. Nature 296: 34–38

Wrightsman RA, Leon W and Manning JE (1986) Variation in antigenic determinants specific to the infective stage of *Trypanosoma cruzi*. Infect Immun 53: 235–239

Immune Responses Against Myelin Basic Protein and/or Galactocerebroside Cross-React with Viruses: Implications for Demyelinating Disease

R. S. FUJINAMI

1   Introduction   93
2   Experimentally Induced Autoimmune Disease and Virus   94
3   Molecular Mimicry and Viruses   95
3.1 Cross-Reactions   95
3.2 Humoral Immune Responses That Cross-React with CNS Components   95
3.3 Cellular Immune Responses That Cross-React with CNS Components   96
3.4 Cross-Reaction with Galactocerebroside-like Moieties   96
4   Implications   97
4.1 Autoimmune Demyelination   97
4.2 Molecular Mimicry Between Virus and Disease-Inducing Sites   98
4.3 Persistence and Immune Recognition   98

References   99

1 Introduction

It has long been speculated that viruses share cross-reacting determinants with their host self-components. Not until recently could this be directly demonstrated by using monoclonal antibodies (Mabs) and by making direct protein sequence comparisons. This concept of shared common determinants has been the basis for understanding the mechanisms of virus-induced autoimmunity.

The encephalomyelopathies following virus infection and other human demyelinating diseases, such as multiple sclerosis, may be explained by the generation of cross-reacting immune responses between viruses and central nervous system (CNS) components. These immune responses would be responsible for the histopathologic and clinical changes observed in individuals with these diseases (FUJINAMI and OLDSTONE, 1986, 1987) and typical of encephalopathies that follow viral infections, particularly infection with measles virus. However, virus is rarely recovered from the CNS, and its presence does not correlate with occurrence of disease (MILLER et al. 1956; AARLI 1974; APPLEBAUM et al. 1949); thus direct, virally caused tissue destruction is unlikely.

The features of this post-infectious encephalomyelitis are consistent with those observed with a CNS disease in the late 1800s and early 1900s after rabies

Department of Pathology, University of California, San Diego, La Jolla, California 92093, USA

vaccination (APPLEBAUM et al. 1953; BLATT and LEPPER 1953; HORACK 1930; REMLINGER 1928; UCHIMURA and SHIRAKI 1957). At that time, the rabies virus vaccination regimen consisted of a dozen or so subcutaneous injections of an attenuated virus preparation grown in rabbits and consisting of dried infected spinal cord (PATERSON 1971). A paralytic disease that involved perivascular infiltrates and motor difficulties was occasionally, yet consistently associated with vaccination. Initially, these instances of neurologic manifestations were attributed to the attenuated virus itself. However, other components in the vaccine became suspect since the attenuated virus preparation was treated with phenol to inactivate any infectivity. The incidence of disseminated encephalomyelitis was similar to that reported before using this phenolic preparation. It was the studies of RIVERS et al. (1933; RIVERS and SCHWENTKER 1935) which unequivocally determined that injection of normal CNS tissue into animals could lead to the development of a disseminated encephalomyelitis closely relating the disease seen in humans after rabies vaccination. In addition, individuals with paralytic disease after rabies virus vaccination had antibodies that reacted with brain tissue (KOPROWSKI and LEBELL 1950), indicating reactivity to CNS components. When CNS tissue emulsified in Freund's complete adjuvant was used for immunization of experimental animals, an acute disease could be reproducibly produced (reviewed in WAKSMAN 1962).

2 Experimentally Induced Autoimmune Disease and Virus

The autoimmune disease known as experimental allergic encephalomyelitis (EAE) is an outgrowth of these early studies. This disease is produced by injection of neuro-antigens with adjuvant into a suitable animal species. Various CNS proteins including myelin basic protein have now been sequenced and regions identified that contain encephalitogenic determinants, i.e., peptide stretches capable of inducing EAE (reviewed in ALVORD 1984). Clinical signs consisting of ataxia leading to paralysis and histological lesions made up of perivascular cellular infiltrates of mononuclear cells are prominent features of the disease. EAE is T-cell mediated (WEIGLE 1980), although CNS-specific antibodies can play a role in the demyelinating lesions (LININGTON et al. 1988; SCHLUESENER et al. 1987; BROSNAN et al. 1983).

Viral infections modulate the expression of EAE. For example, experimentally induced viral infections can exacerbate or increase the incidence of EAE. HOCHBERG et al. (1977) were able to reinduce EAE in rabbits after herpes simplex virus type 1 infection. Both the subcutaneous and intracerebral routes of infection were found to be effective in initiating the disease. In these studies, approximately half the animals given two injections of herpes simplex virus had pathologic evidence of recrudescent EAE. The role of herpes virus in exacerbating the autoimmune disease was unclear; however, these investigators felt that the observed pathology was not herpes virus encephalitis for several reasons. First, the pathology of EAE and that of herpes simplex virus encephalitis are clearly distinct; second, the subcutaneous route of inoculation potentiated the disease; and third, herpes simplex virus could not be cultured from the brains of animals with the recurrent EAE.

A second model has been described by MASSANARI et al. (1979), who potentiated

EAE by persistently infecting hamsters with measles virus. The animals with persistent measles virus infection had a higher incidence of EAE compared with uninfected EAE-sensitized littermates. In this instance, the measles virus shortened the latency period and potentiated the EAE. This EAE was not augmented if the virus was heat-inactivated prior to use, suggesting that live virus was needed for the potentiation. More recently, MOKHTARIAN and SWOVELAND (1987) demonstrated similar results in EAE-resistant mice infected with semliki forest virus. These authors suggested that virus infection facilitated priming or clonal expansion of myelin-reactive cells. However, alterations in the blood-brain barrier, increases in lymphocyte function, or cross-reactivity between virus and self could be reasonable explanations.

3 Molecular Mimicry and Viruses

3.1 Cross-Reactions

Viruses have been demonstrated to have cross-reacting determinants with self-components. Often during a virus infection, antibodies to self-proteins can be detected in the circulation. FUJINAMI, OLDSTONE, and colleagues (FUJINAMI et al. 1983) have shown cross-reactions between measles virus and an intermediate filament protein, cytokeratin; with DALES (DALES et al. 1983) they described a Mab reacted with vaccinia virus hemagglutinin and vimentin, assayed by using immunofluorescent staining techniques and Western blotting analysis. Other cross-reactions with various viruses and self-components are reviewed by FUJINAMI and OLDSTONE (1987).

3.2 Humoral Immune Responses That Cross-React with CNS Components

GOSWAMI et al. (1985) has described a neutralizing Mab to simian virus 5 hemagglutinin-neuraminadase protein that, by immunofluorescence, reacts with Purkinje cells from the brain. This cross-reactivity was further documented using ELISA and RIA with human brain as the antigen. Similarly, KENNEDY et al. (1987) have reported that a cultured human oligodendroglioma cell line and herpes simplex virus-infected cells share common antigenic determinants. This was shown by using polyclonal antibodies and a Mab to herpes virus, both of which reacted with 1 of 60 brain tumor-derived cell cultures. The cross-reacting determinant could be detected by immunofluorescent staining of the oligodendroglioma cell line. In another system, FRIEDMAN et al. (1987) isolated immune complexes from the sera of patients with multiple sclerosis. These immune complexes were then injected into rabbits to raise antisera. Using immunoelectrophoresis, ELISA, and Western blot analysis, the authors found that the antisera reacted with brain proteins from normal persons and multiple sclerosis patients. Interestingly, the antisera immunoprecipitated the nucleocapsid protein of measles virus, suggesting a cross-reacting determinant between brain components and measles virus.

3.3 Cellular Immune Responses That Cross-React with CNS Components

JOHNSON et al. (1984) studied 19 patients with encephalomyelitis following measles virus infection. Their results support the contention that this disease has a pathogenesis similar to that of EAE. In the majority of the individuals tested, a positive lymphocyte proliferative response to myelin basic protein was demonstrable. In addition, some patients with encephalomyelitis following rabies vaccination, varicella, or rubella virus infection had similar clinical and histologic findings. These data suggest a common pathway for the development of demyelinating encephalomyelitis following virus infection resulting in sensitization to CNS tissue. More recently, MATOSSIAN-ROGERS et al. (1987) generated cytotoxic T lymphocytes to measles virus. Killing by these cytotoxic T cells could be enhanced two- to threefold when these T cells were cultured with myelin basic protein prior to performing the killing assay. In addition, these measles virus-specific T lymphocytes have the ability to kill target cells coated overnight with myelin basic protein. Conversely, when cytotoxic T cells were generated to myelin basic protein, these T cells could kill myelin basic protein-coated cells and measles virus-infected target cells. The killing of allogeneic target cells by allo-cytotoxic T cells was not altered by the addition of myelin basic protein. These results suggest a functional cross-reaction with myelin basic protein and measles virus.

3.4 Cross-Reaction with Galactocerebroside-like Moieties

Lately, I have been studying cross-reacting immune responses particularly to CNS determinants and viral epitopes (FUJINAMI and OLDSTONE 1986). The generation of such cross-reacting immune responses could contribute to the demyelinating pattern of disease observed in some of the post-infectious encephalomyelopathies and/or chronic CNS diseases such as multiple sclerosis. One model of human demyelinating diseases is Theiler's murine encephalomyelitis virus (TMEV) infection of mice. TMEV is a murine picornavirus that, when injected intracerebrally into the appropriate strain of mouse, can induce a chronic demyelinating disease (LIPTON et al. 1986). This disease has some histologic features in common with EAE, including perivascular cellular infiltrates and demyelination. The demyelinating disease has been suggested to be immunologically mediated, although the contributions of direct viral-induced pathology have not been resolved.

Recently, one Mab to TMEV was found to react with the virus and galactocerebroside (GC) (FUJINAMI et al. 1988b). GC is a major component of myelin. This was demonstrated in an initial screening assay by a solid phase ELISA using a TMEV antigen preparation of GC. This Mab, H8, was found to react very strongly with the TMEV antigen preparation and less avidly but consistently with the GC preparation. To characterize Mab H8 further, Western blot analysis was conducted to determine two features of the binding: was a linear determinant recognized, and what viral protein was bound or specified? Purified virus was used as the antigen, and viral proteins were separated by SDS-PAGE. Binding was established by Western blot analysis and visualized by autoradiography. The fact that Mab H8 bound to vp-1, the outermost viral capsid protein, suggests the recognition of a linear antigenic determinant. To determine whether the Mab reacts with a neutralizing site in vp-1, viral plaque reduction titration assays were performed. The result was that Mab H8 efficiently neutralized TMEV. Therefore, it defines a neutralizing site on vp-1.

Experiments were then initiated to determine whether TMEV could be neutralized using antibodies to GC. Two Mabs to GC (generated during TMEV infection) were tested, in addition to Mab H8, as well as another Mab to GC, obtained from Dr. Hilary Koprowski at the Wistar Institute. The results show that H8 neutralized TMEV very efficiently. The other three Mabs to GC also had the ability to neutralize TMEV but somewhat less efficiently. Therefore, one of the neutralizing sites on TMEV involves a site that cross-reacts with or is in close proximity to GC or a lipidlike structure associated with vp-1.

To determine whether such antibodies exist during viral infection (i.e., antibodies to GC) mice were infected with TMEV and sera were collected at various times postinfection. TMEV-specific antibodies were detected within 7 days postinfection. Further, antibodies to GC were detectable approximately 10 days after infection by ELISA. Interestingly, antibodies to myelin basic protein, another CNS antigen, were also present. Therefore, during viral infection involving CNS demyelination, antibodies to viral and CNS components were present.

Since antibodies to GC were found during the infection of mice with TMEV, studies were initiated to determine whether the cross-reacting Mab could react with nerve tissue. Brain cells from newborn mice were established in cultures on coverslips and harvested at a time when various CNS cell populations were present, i.e., astrocytes and oligodendrocytes. The coverslops were then stained with the Mab H8. By immunofluorescent staining, the Mab, could be seen to react with a cell population consistent with oligodendrocyte-like cells in the cultures. These cells had extended processes and grew on top of the majority of glial fibrillary acidic protein-positive cells. In other studies with Dr. Henry Powell at the University of California, San Diego, Mab H8 was injected into the right sciatic nerve of rats, and the left was injected with control Mab. In half the animals (3/6), the right sciatic nerve showed Schwann cell destruction and demyelination. No similar changes were observed in the left sciatic nerve (unpublished data). These data suggest that the antibody binds to myelin structures and plays a role in the demyelinating disease.

4 Implications

4.1 Autoimmune Demyelination

A cross-reacting immune response initially generated against a virus that also reacts with CNS components could lead to demyelination. The virus may be eliminated by the early immune response, yet a chronic immune response that cross-reacts with CNS tissue could continue. This chronic immune response would be originally driven by aberrant expression of class II and/or class I antigens induced by virus infection of the target tissue and later by the presence of inflammatory cells and a self-CNS component such as myelin basic protein. Thus, trying to identify the inciting agent after the fact may prove fruitless.

Virus infection of the CNS could lead to tissue injury and limited pathology. However, by superimposing an immune response against a cross-reacting determinant such as GC, a more extensive pattern of disease may result. For instance, BROSNAN et al. (1977) described a model in which a delayed type hypersensitivity response is generated within the eyes of experimental host rabbits. No demyelination is observed under these conditions, yet a substantial inflammatory response is present. If antibody to myelin or GC is injected into the rabbit eye, no pathologic changes are observed. However, when an inflammatory response is generated and CNS-specific antibodies are present simultaneously, demyelination is clearly demonstrable. Thus, the cross-reacting immune response may increase the initial damage caused by virus infection of the CNS.

4.2 Molecular Mimicry Between Virus and Disease-Inducing Sites

The prediction is that the common determinant between a virus protein or antigenic determinant and a host CNS epitope must occur in disease-inducing regions, such as the brain encephalitogenic region of myelin basic protein. This region would be capable of initiating and sustaining a disease like EAE in a genetically suitable host. An example of such a model was described by FUJINAMI and OLDSTONE (1986), who found that the encephalitogenic region of rabbits shares six amino acids in common with the hepatitis B virus polymerase. Immunization with the viral peptide induced autoantibodies in most of the rabbits, cellular reactivity to myelin basic protein in half, and histologic disease in some. It is predicted that an immune response against a nondisease-inducing region could induce autoantibodies, but no disease would be observed.

4.3 Persistence and Immune Recognition

In addition to the possibility of inducing a cross-reacting immune response that results in autoimmunity, it may be advantageous for a virus to share common determinants with its host's self-components. First, viruses are assembled and mature in discrete sites within the infected cell, often in association with cellular structures such as intermediate filament proteins. By having common structures or signal regions incorporated into the viral proteins, these proteins could be transported to sites within the cell for viral assembly to take place. For instance, paramyxoviruses are assembled in the cytoplasm of infected cells in association with intermediate filament proteins. By having regions in common with these cellular proteins, the viral proteins may be brought to cellular compartments where the intermediate filament proteins reside.

Another possibility for the evolution of common regions between viral and cellular proteins would be that, if virus looked enough like self, an effective immune response would not be generated. Thus, a virus could establish a persistent infection rather than be cleared by the ensuing antiviral immune response. We have described such a situation with human cytomegalovirus in which there is a common region between the virus immediate early 2 region and HLA DR-beta chain (FUJINAMI et al. 1988a).

HLA DR antigens are involved in immune recognition, and aberrant expression of DR antigens could lead to modulation of the immune response.

Acknowledgements. This work was supported in part by The National Institutes of Health, NS 23162, and The National Multiple Sclerosis Society, GR 1780-A. I would like to thank Peggy Farness and Susan McClanahan for their technical support and Jan Richards for manuscript preparation.

References

Aarli JA (1974) Nervous complications of measles: cinical manifestations and prognosis. Eur Neurol 12: 79–93

Alvord EC, Jr (1984) Species-restricted encephalitogenic determinants. In: Alvord EC, Kies MW, Suckling AJ (eds) Experimental allergic encephalomyelitis. Liss, New York pp 523–537

Applebaum E, Dolgopol VB, Dolgin J (1949) Measles encephalitis. Am J Dis Child 77: 24–48

Applebaum E, Greenberg M and Nelson J (1953) Neurological complications following antirabies vaccination. J Am Med Assoc 151: 188–191

Blatt NH and Lepper MH (1953) Reactions following antirabies prophylaxis. Report on 16 patients. Am J Dis Child 86: 395–402

Brosnan CF, Stoner GL, Bloom BR, Wisniewski HM (1977) Studies on demyelination by activated lymphocytes in the rabbit eye II. Antibody-dependent cell mediated demyelination. J Immunol 118: 2103–2110

Brosnan CF, Traugott U, Raine CS (1983) Analysis of humoral and cellular events and the role of lipid during CNS demyelination. Acta Neuropathol (Berl) [Suppl] 9: 59–70

Dales S, Fujinami RS, Oldstone MBA (1983) Infection with vaccinia favors the selection of hybridomas synthesizing autoantibodies against intermediate filaments, one of them cross-reacting with the virus hemagglutinin. J Immunol 131: 1332

Friedman J, Buskirk D, Marino LJ, Zabriskie JB (1987) The detection of brain antigens within the circulating immune complexes of patients with multiple sclerosis. J Neuroimmunol 14: 1–17

Fujinami RS, Oldstone MBA (1986) Amino acid homology and cross-reacting immune responses between the encephalitogenic site of myelin basic protein and virus: virus-induced autoimmunity. In: Brown F, Chancok RM, Lerner RA (eds) Vaccines 86: modern approaches to immunizations. Cold Spring Harbor Laboratory, New York, pp 247–251

Fujinami RS, Oldstone MBA (1987) Molecular mimicry as a mechanism for virus induced autoimmunity. In: Progress in Allergy, Triggering Factors in Autoimmunity, S Karger AG Basel

Fujinami RS, Nelson JA, Walker L, Oldstone MBA (1988a) Sequence homology and immunologic cross-reactivity of human cytomegalovirus with HLA-DR beta chain: a means for graft rejection and immunosuppression. J Virol 62: 1001–1005

Fujinami RS, Zurbriggen A, Powell HC (1988b) Monoclonal antibody defines determinant between Theiler's virus and galactocerebroside. J Neuroimmunol 20: 25–32

Goswami KKA, Morris RJ, Rastogi LS, Lange LS, Russell WC (1985) A neutralizing monoclonal antibody against a paramyxovirus reacts with a brain antigen. J Neuroimmunol 9: 99–108

Hochberg FH, Lehrich JR, Arnason GW (1977) Herpes simplex infection and experimental allergic encephalomyelitis: an experimental model system for reactivation of EAE. Neurology 27: 584–587

Horack HM (1930) Allergy as a factor in the development of reactions to anti-rabic treatment. Am J Med Sci 197: 672–682

Johnson RT, Griffin DE, Hirsch RL, Wolinsky JS, Roedenbeck S, Soziano IL, Vaisberg A (1984) Measles encephalomyelitis–clinical and immunologic studies. N Engl J Med 310: 137–141

Kennedy PGE, Watkins BA, LaThangue NB, Clements GB, Thomas DGT (1987) A cultured human oligodendroglioma cell line and herpes simplex virus-infected cells share antigenic determinants. J Neuro-Oncology 4: 389–396

Koprowski H, LaBelle I (1950) The presence of complement-fixing antibodies against brain tissue in sera of persons who had received antirabies. Am J Hyg 51: 292–299

Linington C, Bradl M, Lassmann H, Brunner C, Vass K (1988) Augmentation of demyelination in rat acute allergic encephalomyelitis by circulating mouse monochlonal antibodies directed against a myelin/oligodendrocyte glycoprotein. Am J Pathol 130: 443–454

Lipton H, Miller S, Melvold R, Fujinami RS (1986) Theiler's murine encephalomyelitis virus (TMEV) infection in mice as a model for multiple sclerosis. In: Notkins AL, Oldstone MBA (eds) Concepts in viral pathogenesis II Springer-Verlag, Berlin Heidelberg New York

Massanari RM, Paterson PY, Lipton HL (1979) Potentiation of experimental allergic encephalomyelitis in hamsters with persistent encephalitis due to measles virus. J Infect Dis 139: 297–303

Matossian-Rogers A, Dos Santos A, Festenstein H (1987) Human cytotoxic T-cells against measles virus-infected and myelin basic protein-coated targets are cross-reactive. Int Arch Allergy Immunol 84: 159–164

Miller HG, Stanton JB, Gibbons JL (1956) Para-infectious encephalomyelitis and related syndromes: a critical review of the neurological complications of certain specific fevers. Q J Med 25: 427–505

Mokhtarian F, Swoveland P (1987) Predisposition to EAE induction in resistant mice by prior infection with semliki forest virus. J Immunol 138: 3264–3268

Paterson PY (1971) The demyelinating diseases: clinical and experimental correlates. In: (eds) Little, Brown, New York, pp 1269–1300

Remlinger P (1928) Les paralysies du traitement anti-rabique. Ann Inst Past 42: 71–132

Rivers TM, Schwentker FF (1935) Encephalomyelitis accompanied by myelin destruction experimentally produced in monkeys. J Exp Med 61: 689–702

Rivers TM, Sprunt DH, Berry GP (1933) Observations on attempts to produce acute disseminated encephalomyelitis in monkeys. J Exp Med 58: 39–53

Schluesener HJ, Sobel RA, Linington C, Weiner HL (1987) A monoclonal antibody against a myelin oligodendrocyte glycoprotein induces relapses and demyelination in central nervous system autoimmune disease. J Immunol 139: 4016–4021

Uchimura I, Shiraki H (1957) A contribution to the classification and the pathogenesis of demyelinating encephalomyelitis, with special reference to the central nervous system lesions caused by preventive inoculation against rabies. J Neuropathol 16: 139–203

Waksman BH (1962) Auto-immunization and the lesions of auto-immunity. Med 41: 93–141

Weigle WO (1980) Analysis of autoimmunity through experimental models of thyroiditis and allergic encephalomyelitis. Adv Immunol 30: 159–273

Molecular Mimicry: Parasite Evasion and Host Defense

R. T. Damian

1 Introduction 101
2 What Is and What Is Not Molecular Mimicry 102
3 Direct Evidence for Molecular Mimicry in Parasites 104
4 Parasite Mimicry of Host Receptors and/or Ligands 104
5 Eclipsed Epitopes 107
6 Indirect Evidence for Molecular Mimicry 108
7 How May Molecular Mimicry Favor Parasitism? 109
8 Conclusions 110
References 111

1 Introduction

Molecular mimicry was formally defined as early as 25 years ago (DAMIAN 1964) as the sharing of antigenic determinants between parasite and host. The possible consequences of molecular mimicry to the host-parasite relationship were then thought to be basically threefold: autoimmunity, parasite avoidance of host immunity, and the development of host antigenic polymorphisms. One focus of my 1964 paper, in common with the contributions of others (SPRENT 1962; DINEEN 1963), was on the second possibility, that of parasites coexisting with their hosts by means of their failure to stimulate immunity. Earlier, I had introduced the term "eclipsed antigen" for a mimetic parasite epitope that goes unrecognized as foreign in the host (DAMIAN 1962). The idea to be conveyed was that a parasite's epitope could be "hidden" behind the corresponding host epitope from the host's immune recognition system. Although this term has not caught on to the extent that "molecular mimicry" has, I still believe it to be a potentially useful one, especially if modified to include more modern terminology: thus, "eclipsed epitope" or "eclipsed mimetic epitope". A mimetic epitope may operate in immune evasion by modulating immunoregulation in the parasite's favor (DAMIAN 1984; GROSSMAN et al. 1986; KEMP et al. 1986), e.g., as may occur in Fc receptor mimicry by schistosomes (TORPIER et al. 1979; TARLETON and KEMP 1981) or by other mechanisms. In this case, it may be

Department of Zoology, University of Georgia, Athens, GA 30602, USA useful to call it an "aggressive epitope" or "aggressive mimetic epitope", following the ideas on aggressive mimicry developed by WICKLER (1968) and VANE-WRIGHT (1976), and playing a bit loosely with the use of the term "epitope." If a mimetic epitope provokes autoimmunity, then perhaps it could be referred to as an "autoimmunogenic epitope."

I also advanced a hypothesis in that paper (DAMIAN 1964) for the third possibility, which was that molecular mimicry generates considerable selection pressure on host populations for the development and maintenance of antigenic polymorphisms, such as the blood group systems. HALDANE (1949) foreshadowed this idea by suggesting that it would be advantageous against pathogens for an individual to possess a rare biochemical phenotype. He further wrote, "because of its rarity it will be resistant to diseases which attack the majority of its fellows. And it means that it is an advantage to a species to be biochemically diverse, and even to be mutable as regards genes concerned in disease resistance." The role of molecular mimicry in host antigenic polymorphisms was further discussed by DAMIAN (1979).

Autoimmunity as a consequence of molecular mimicry had been promulgated 2 years earlier by ROWLEY and JENKIN (1962), while workers in other systems were already considering this possibility (MARKOWITZ et al. 1960; KAPLAN and MEYESERIAN 1962). Molecular mimicry-mediated autoimmunity is now considered to be an important mechanism of viral pathogenesis (OLDSTONE 1987). The current focus has been fueled by the surge of new data on shared host-parasite epitopes and peptide sequences deriving from the use of monoclonal antibodies (SRINIVASAPPA et al. 1986) and sequence homology computer searches (JAHNKE et al. 1985; FUJINAMI and OLDSTONE 1985; MC LAUGHLIN et al. 1987). This new evidence suggests that molecular mimicry may be even more common than was originally suspected. The experimental demonstration of tissue injury resulting from cross-reacting host-viral epitopes (FUJINAMI and OLDSTONE 1985) has also provided a fillip to research in this field.

It is the purpose of the present paper to review and examine molecular mimicry in parasites, its roles in parasite evasion of host immunity and other aspects of parasitism, and host defenses against it. Since I (DAMIAN 1979) and others (FEDOTOVA and GROMOVA 1974; DeVAY and ADLER 1976; YOSHINO and BOSWELL 1986; BAYNE et al. 1987) have considered molecular mimicry by zoo- and phytoparasites of invertebrate animals and plants, this review will be virtually confined to the phenomenon as it exists in protozoan and helminthic endoparasites of vertebrate animals, e.g., in "traditional" parasites of hosts with the most highly evolved immune responses.

2 What Is and What Is Not Molecular Mimicry

When the term "molecular mimicry" was coined (DAMIAN 1964), it referred to antigenic mimicry, and the functionality of host-parasite molecular sharing was stressed by discussing its possible consequences. I later (DAMIAN 1979) attempted to broaden the use of "molecular mimicry" to include other biologically relevant, functional relationships involving similar or shared molecular structures, a subject recently developed in great detail by STOWE (1988). Thus, molecular mimicry is not fundamentally different from other forms of biological mimicry. Some selective advantage for one or both partners of a symbiotic (*sensu lato*) or ecological association must occur to stabilize the relationship.

Since molecular similarity is a reflection of phylogeny, close phylogenetic affinity would overshadow molecular mimicry. When dealing with hosts and parasites, this concern is somewhat mimimized since hosts and parasites are, at least medically speaking, phylogenetically disparate. Except for the fetus, mammalian parasites of mammals are unknown. But the evolutionary conservation of molecules fundamental to life spans even prokaryotes and eukaryotes, a fact so well-known as to make repeating it almost trite. Heat-shock proteins (hsp) provide an excellent example. They have been found among parasites in trypanosomatids (VAN DER PLOEG et al. 1985; GLASS et al. 1986; DRAGON et al. 1987), *Plasmodium* (BIANCO et al. 1986; YANG et al. 1987; ARDESHIR et al. 1987), *Schistosoma* (NENE et al. 1986; HEDSTROM et al. 1987; YUCKENBERG et al. 1987), and the filarid, *Brugia malayi* (SELKIRK et al. 1987). The ubiquity and high degree of conservation of hsps argue for their basic importance in cells (LINDQUIST 1986). Among other functions, new roles for them in intracellular protein translocations are rapidly being described (CHAPPELL et al. 1986; DESHAIES et al. 1988; CHIRICO et al. 1988). For a special role in parasitism, perhaps one need look no farther than the heat-shock response itself for an explanation of their adaptive value for parasites. This may be true at least for the case of the parasites listed above, all of which alternate between homiothermic and poikilothermic hosts in their life cycles. The heat-shock response could easily provide the developmental triggers needed for them to adapt to the vastly different environments provided by two physiologically dissimilar hosts (VAN DER PLOEG et al. 1985). Yet one should not be too quick to dismiss the possibility of molecular mimicry at work, *superimposed* upon a deeper physiological significance for the conserved, shared molecular structures, in the special relationship that is parasitism. Again the hsps are instructive. POLLA (1988) has hypothesized a role for them in inflammation, based upon their induction by a number of stimuli besides the heat shock of fever, and the existence of many links between immune mediators and the heat-shock response. She contemplated whether hsps would favor host or parasite in their relationship and suggested that parasites might employ molecular mimicry to abuse the host by means of hsp autoimmunogenic epitopes, although the advantage of this strategy to a parasite was not made clear. YOUNG et al. (1988) have also suggested that hsps may contain autoimmunogenic epitopes. On the other hand, antigenicity of conserved hsps may be a mechanism of host defense through cross-immunization against a variety of pathogens (YOUNG et al. 1988), and they may also be involved in nonreciprocal cross-immunity (SCHAD 1966), a possible mechanism for interparasite competition mediated by the host immune response.

Perhaps parasite hsps are involved more directly in immune evasion. MUNRO and PELHAM (1986) have established the identity of a mammalian hsp 70-like protein with the immunoglobulin heavy chain binding protein or "BiP" (HAAS and WABL 1983). This suggests that certain parasite hsps may also have affinity for heavy chains. If they are expressed on their surface under stress, they could be involved in the non-Fab binding of antibodies known to take place in parasite surfaces, as for example, on schistosomes (KEMP et al. 1986). These authors discuss various ways by which "flagrant" binding of host immunoglobulins could favor parasite survival, including provocative suggestion that the idiotype regulation network may be altered, modifying the immune response to favor parasite survival.

Parasites might capitalize upon molecular conservation to minimize their foreignness to hosts. For example, actin, a highly conserved protein, is used by *Schistosoma mansoni* to construct its surface spines (COHEN et al. 1982; DAVIS et al. 1985; MATSUMOTO et al. 1988). These spines, prominent on the male worm, have the obvious function of anchoring the worm pair in the venule for oviposition by the female worm. The actin of the spines is probably also mobilized and used for repair of damaged areas on the surface (MCCORMICK and DAMIAN 1987; MATSUMOTO et al. 1988). In addition, some hsps are able to cross-link actin filaments (KOYASU et al. 1986). MCCORMICK and DAMIAN (1987) have discussed the possible interrelationships among complement-mediated damage, calcium flux, actin, actin cross-linking proteins, and surface repair in the schistosomes. Constraints imposed upon the parasite by the ever-present host immune response could thus have played a role in the choice of molecules for parasite organelle construction. In this way, molecular mimicry could have become superimposed upon the phylogenetic conservation of molecular structures.

Finding similar molecular structures in parasite and host is but the first step in any investigation of molecular mimicry. In practice, it is common to refer to this first step as molecular mimicry. This usage is acceptable, since functionality is an implied, if not always stated, working hypothesis, deriving from the parasitic relationship itself. In recent years, molecular mimicry has come to denote engineered as well as biological mimicry at the molecular level (LINTHICUM and BOLGER 1985; SACKS et al. 1985; LINTHICUM and NADIR 1988; LERNMARK et al. 1989).

3 Direct Evidence for Molecular Mimicry in Parasites

The original evidence upon which the molecular mimicry hypothesis was based was the serological demonstration of host antigen in parasite material, usually either by inhibition of specific erythrocyte agglutination or by the stimulation — through immunization or infection — of antibodies that cross-react with host antigens (DAMIAN 1964). The importance of differentiating true molecular mimicry and host antigenic contamination was stressed. Most prominent among the host-parasite cross-reacting antigens known at the time were the blood group antigens, but many others were also soon described (CAPRON et al. 1965, 1968; DAMIAN 1967).

Although this type of evidence continues to be gathered, newer approaches are less ambiguous with respect to possible contamination by host material. These include the use of in vitro labeling of proteins and saccharide structures, use of monoclonal antibodies, use of cloned parasite genes and deduced peptide sequences, use of synthetic peptide probes (DYRBERG and OLDSTONE 1986), and direct structural, comparative studies of parasite and host molecules.

4 Parasite Mimicry of Host Receptors and/or Ligands

Attachment to and invasion of host cells are critical steps for successful parasitism by many medically important protozoa, including malaria organisms and trypanosomatid flagellates. It is well-established that these processes are mediated by specific cell surface ligands (reviewed by HADLEY et al. 1986). CHANG and FONG (1983) suggested that, through molecular mimicry, leishmanial parasites may take advantage of functional macrophage receptors to gain entry into these cells. Much evidence has recently accrued in support of the hypothesis that molecular mimicry of peptides and oligosaccharides plays a prominent role in intracellular parasitism. Since the intracellular environment is an immunologically privileged site, receptor or ligand mimicry by which a parasite can enter this environment may properly be considered a form of immune escape. In addition, the adaptive value of molecular mimicry to intracellular parasites would transcend immunity, since their life needs are provided for by the intracellular milieu.

Insofar as peptides are concerned, most of the evidence for receptor and/or ligand mimicry comes from the molecular biology of the malaria parasites, genus *Plasmodium*. A considerable sequence homology exists between the conserved part of variable region II of the circumsporozoite (CS) protein (CSP) of *Plasmodium* (DAME et al. 1984) and region E of thrombospondin (TSP), a platelet-derived adhesive glycoprotein (DIXIT et al. 1986; LAWLER and HYNES 1986). Other malaria proteins also have homology regions with a variety of host proteins. CHEUNG et al. (1986) found sequence homology between a *P. falciparum* merozoite surface antigen and the nonhelical parts of host intermediate filament proteins. HOLMQUIST et al. (1988) reported a short homology stretch between *P. falciparum* ring-infected surface antigen Pf155/RESA, an analogous antigen in *P. chabaudi* (Pch105), and erythrocyte band 3 protein from both humans and mice. McLAUGHLIN et al. (1987) compared published, tandemly repeated, immunogenic, amino acid sequences of *Plasmodium* proteins with human and human viral sequences in a computer database. They found a total of 29 matches (of at least four amino acids) with human proteins and 26 matches with human viruses among the six parasite sequences examined. Many of these shared sequences may be too short to be functional epitopes. Another *P. falciparum* vaccine candidate proteins, Pfs25, contains four tandemly-repeated epidermal growth factor (EGF)-like domains (KASLOW et al. 1988).

These sequence homologies have been suggested by their discoverers to favor parasitism functionally. The most obvious way would be through receptor or ligand mimicry. An excellent case for this mode can be made for the sequence homology between CSP and TSP. TSP has been suggested to mediate cytoadherence between infected erythrocytes and vascular endothelium and could thus contribute critically to the congestive pathology of *falciparum* malaria (ROBERTS et al. 1985). Indeed, recent evidence suggests that infected erythrocytes contain a parasite protein, "thrombospondin-related anonymous protein" or TRAP, which may directly or indirectly enable the red cell surface to bind to TSP receptors (ROBSON et al. 1988). But more to the point of this discussion, hepatocytes, which are the cells initially infected by the mosquito-injected sporozoite stage of the parasite, bind sporozoites through receptors for CSP, and TSP region E has affinity for liver cells. These facts led FRAZIER (1987) to speculate that if hepatocytes actually bind CSP through a region II receptor, "then the receptor in liver may be designed to bind TSP by recognizing its type E repeats." If this were in fact the case, it would represent another instance of the appropriation of a host recognition system to gain entry, in this case, to the liver. The TRAP protein, which occurs in the blood stages of *P. falciparum*, shares a sequence (WSPCSVTCG) with CSP as well as with TSP (ROBSON et al. 1988; GOUNDIS and

REID 1988), and this peptide may therefore function in the sporozoite invasion of hepatocytes. Similarly, the EGF-like sequences of Pfs25 have been suggested to play a role in ookinete invasion of mosquito epithelial cells (KASLOW et al. 1988). MCLAUGHLIN et al. (1987) considered the biological significance of the many sequence matches they found between *Plasmodium* proteins and human proteins or human virus proteins to be perhaps some form of immune escape (including possibly molecular mimicry through tolerance). Interestingly, most of the matches were to proteins of blood or to proteins of leukocytotropic viruses, which if not biased by over-representation in the existing database, lends some credence to the idea.

Molecular mimicry may function in cell invasion in other ways besides receptor/ligand mimicry. With respect to their finding of mimicry between the merozoite surface and the host cell cytoskeleton, CHEUNG et al. (1986) speak of a "camouflage hypothesis" whereby "the merozoite surface has evolved to integrate into the host-cell structure during replication of the parasite." Another role with respect to the host cell cytoskeleton was suggested to explain RESA-band 3 mimicry by ANDERS et al. (1988). They hypothesized that since band 3 interacts indirectly with cytoskeleton proteins, then parasite mimetic molecules could disrupt the interactions, thus increasing the "fluidity of the lipid bilayer, which presumably is necessary for the formation of the parasitophorous vacuole at the time of merozoite invasion."

There is considerable concern among the workers in this field that these shared sequences, some of them such promising targets for receptor blockade by vaccine-induced immunity, may be potential autoimmunogenic epitopes (MCLAUGHLIN et al. 1987; KASLOW et al. 1988).

Trypanosomatid parasites utilize a peptide receptor for host cell adherence, the Arg-Gly-Asp (RGD) sequence (WYLER et al. 1985; OUAISSI et al. 1986) found in fibronectin (FN) and many other host proteins including TSP (RUOSLAHTI and PIERSCHBACHER 1987). Trichomonad protozoans also apparently mimic an FN binding site (RIBAUX et al. 1983), and the recently described TRAP protein of *P. falciparum* blood stages has an RGD sequence as well (ROBSON et al. 1988). As discussed by OUAISSI (1988), binding may be effected in two ways: through the parasite's attachment by FN receptors to an unoccupied RGD site on cell-bound dimeric FN, or through molecular mimicry of FN itself. Mimicry of FN may also occur in the helminth parasite, *Schistosoma mansoni* (OUAISSI et al. 1984). Cytoadherence apparently does not play a role in this parasite's relationship with its host, as it does so prominently for the protozoans that display mimicry of FN. Therefore, FN mimicry (if truly molecular mimicry) by the schistosome probably functions in other ways, such as complement binding or worm cytoskeleton organization, as suggested by QUAISSI et al. (1984). Complement receptors have been described from several parasites, including protozoa (JACK and WARD 1980) and helminths (SANTORO et al. 1979; MCGUINNESS and KEMP 1981). Their probable role in intracellular parasitism may relate to host cell invasion, but how they may promote extracellular parasitism is unclear. KEMP et al. (1986) suggested that the C3 receptor of schistosomes may complex with and inactivate potentially damaging C3, perhaps first by its internalization from the exposed parasite surface. Complement receptors, along with the parasite Fc receptors mentioned above, may also qualify as "aggressive epitopes" if their action results in a parasite-favoring perturbation of the immune response (KEMP et al. 1986).

The recent description of sequence homology between *Plasmodium* CS and TRAP proteins and properdin led the discoverers to speculate on its possible adaptive functions, which included use of the CR1 receptor to invade cells and also host tolerance through molecular mimicry (GOUNDIS and REID 1988; ROBSON et al. 1988).

Oligosaccharide groups as well as peptide groups function in parasite adherence and invasion of host cells by the mechanism recently named lectinophagocytosis (OFEK and SHARON 1988). HANDMAN and GODING (1985) demonstrated that a leishmanial glycolipid (L-GL) is involved in binding of the parasite to its host cell, the macrophage, as well as a number of other possible functions (TURCO 1988). Another leishmanial surface glycoconjugate, gp63 (RUSSELL and WILHELM 1986), binds to macrophage complement and mannose-fucose receptors (MOSSER and EDELSON 1985; BLACKWELL 1985). As with peptide ligands, molecular mimicry may be involved in some of these types of interactions. For example, there exist reports of shared carbohydrate epitopes between leishmania parasites and host cells (DECKER-JACKSON and HONIGBERG 1978; GREENBLATT et al. 1984).

5 Eclipsed Epitopes

Among eukaryotic parasites, oligosaccharide epitopes may prove to be more important than peptide epitopes in the "classical" conception of molecular mimicry involving pseudotolerance (DAMIAN 1964; GREENBLATT 1983; YOSHINO and BOSWELL 1986). Major efforts to analyze these structures in parasites are under way.

Our group has undertaken a systematic analysis of the oligosaccharide epitopes synthesized by *S. mansoni*, beginning with those found on glycoproteins. Their glycoproteins contain a variety of *O*- and *N*-linked oligosaccharides (NYAME et al. 1987, 1988a, 1988b, 1989). We recently described the occurrence in these glycoproteins of parasite-synthesized terminal *O*-linked *N*-acetyl-glucosamine residues (NYAME et al. 1987). This structure had previously been found for the first time in cellular proteins isolated from mice (TORRES and HART 1984) and rats (HOLT and HART 1986). Although rodents serve as natural and laboratory hosts for this parasite, this is unlikely to be a case of molecular mimicry. It now appears that this novel protein-saccharide linkage is involved in trafficking of glycoproteins to the nucleopore (FINDLAY et al. 1987). Thus, these shared structures probably represent conservation of fundamentally important eukaryotic biochemistry. *N*-linked chains of the high mannose type were also found in *S. mansoni*. They had a size range of from $Man_7GlcNAc_2$ to $Man_9GlcNAc_2$, which makes them very similar to those synthesized by mammals (NYAME et al. 1988a). The complex type *N*-linked chains of *S. mansoni* also have some similarities to mammalian complex type *N*-linked chains, but they have a number of important differences as well (NYAME et al. 1989). The high mannose class of *N*-linked oligosaccharide chains would constitute better candidates for molecular mimicry (after their immunogenicity is ascertained), and the complex type chains may contain some candidates for anti-schistosome molecular vaccines. An example of a mimicked protective oligosaccharide epitope was provided by DISSOUS et al. (1986). In this case, the epitope is shared with the snail intermediate host.

6 Indirect Evidence for Molecular Mimicry

One form of indirect evidence for functional molecular mimicry derives from putative geographical associations such as those thought to exist between certain blood group antigen-like serotypes of *Leishmania donovani* and *L. braziliensis* and human blood groups (GREENBLATT et al. 1981). This type of evidence is but suggestive, at best.

Another type of indirect evidence consists of classical genetic studies of parasite susceptibility. Just as dominance of immune responsiveness implied the existence of immune response genes (BENACERRAF and McDEVITT 1972), so dominance of *susceptibility* may signify the operation of molecular mimicry in the system (SNELL 1968; GASSER 1969; MARCHALONIS et al. 1974). Unfortunately, this is complicated by the frequent finding of codominance in immunogenetic analyses (EBRINGER et al. 1976). Nevertheless, WRIGHT et al. (1988) found that hybrids between strains of mice (one genetically resistant, the other susceptible to schistosomes) were susceptible to this parasite, leading them to suggest the possibility of molecular mimicry.

On the other hand, dominance of susceptibility may also be interpreted as evidence for the existence of immune suppression genes (SASAZUKI et al. 1983). Indeed, this group has good evidence (HIRAYAMA et al. 1987) that in human schistosomiasis caused by *Schistosoma japonicum*, nonresponsiveness to schistosomal antigen is dominantly inherited, linked to the HLA system, and based upon *Is* genes. MITCHISON and OLIVEIRA (1986) go a step further to hypothesize that the driving force behind the evolutionary development of antigen-specific immune suppression genes and their polymorphisms was and is chronic infection by parasites. In their view, the immunopathology that often accompanies chronic parasitic infection is a strategem used by the parasite, so as to select for immunopathogenesis-controlling *Is* genes which in turn, through their action, prolong the reproductive lifespan of the parasite. Obviously, although not stated, such a mechanism would be facilitated by molecular mimicry. Other authors (DOENHOFF et al. 1985; DAMIAN 1987a) suggest a more direct way in which parasite-induced chronic immunopathogenesis could favor parasite survival. In this view, elaborated to explain certain relationships between the immune response, immunopathogenesis, and schistosome egg excretion from the host, the parasite has harnessed the immunopathogenic response to complete a necessary part of its life cycle, namely, the exteriorization of its eggs into the environment. Examples of related strategies used by other parasites were brought together with the schistosome evidence and named "immune exploitation" (DAMIAN 1987a).

WASSOM et al. (1987) have recently presented specific hypotheses whereby dominant immune suppression genes may operate in murine nematode infections. They observed that inbred strains with defective I-E gene product expression are resistant to both *Trichinella spiralis* and *Nematospiroides dubius*, and that the F_1 hybrids between I-E deficient and I-E normal strains are susceptible. On the basis of these and other findings, they suggest that either I-E-restricted suppressor cells are induced or that I-E-restricted cells induce autoreactive T cells that may suppress protective, I-A-restricted, anti-parasite responses. The latter hypothesis is formally equivalent to that of HIRAYAMA et al. (1987) in that an *Is* gene is epistatic to an *Ir* gene in controlling anti-parasitic immune responses. BLACKWELL et al. (1980) found that, in B10 background mice, cure and noncure phenotypes for experimental visceral leishmaniasis (caused by *Leishmania donovani*) were linked to H-2. In this system, cure is recessive to noncure. Further analysis (BLACKWELL and ROBERTS 1987) using I-A-specific and I-E-specific monoclonal antibodies administered in vivo to *L. donovani*-infected, noncure-phenotype mice showed a pronounced dichotomy of effects. Anti-I-A treatment exacerbated chronic hepatosplenic disease, whereas anti-I-E treatment allowed enhanced late clearance of parasites from the liver and spleen. These authors also hypothesize epistasis between an *Is* and an *Ir* gene.

A third line of indirect evidence for molecular mimicry in parasites is the demonstration in them of enzymes known to be involved in host epitope biosynthesis, particularly oligosaccharide epitopes. RUMJANEK and SMITHERS (1978) and RUMJANEK et al. (1978) have shown that schistosome tegumental carbohydrates were synthesized by a process fundamentally similar to that of their hosts in that glycosyl transferases and lipid acceptors were used. With respect to molecular mimicry, this approach was used by BEN-ISMAIL et al. (1982b) to detect the enzymes directing the final synthesis of the A, H, and Lewis blood group epitopes, namely, 3-a-*N*-acetyl-D-galactosaminyltransferase, 2-a-L-fucosyltransferase, and 4-a-L-fucosyltransferase, in another trematode worm, the sheep liver fluke *Fasciola hepatica*. This group had earlier demonstrated the existence of these epitopes in the fluke by inhibition reactions, immunogenicity, and immunofluorescent localization (BEN-ISMAIL et al. 1982a). Since neither enzymes nor blood group antigens was metabolically labelled in the worms, the possibility remains that either or both could have been acquired from the sheep and cattle from which the worms were obtained. This appears to be unlikely, however, at least for the Lewis specificities, which are absent in these animals. Their results emphasize the potential for molecular mimicry that is inherent in the sharing of glycosylation pathways between eukaryotic animal parasites and their hosts.

7 How May Molecular Mimicry Favor Parasitism?

Some viewpoints on how molecular mimicry could confer selective advantages to parasites have been brought out above and include the idea of receptor or ligand mimicry for cell adherence and entry by intracellular parasites, as well as some ideas relating to the maintenance of host cell integrity during parasite entry or replication.

An older viewpoint of molecular mimicry in parasitism has been to assign to it a role in immune evasion. Immune evasion may take several forms (BLOOM 1979), but the most fundamental division with respect to molecular mimicry is between evasion at the afferent limb versus evasion at the efferent limb of the immune response.

Afferent limb evasion, i.e., nonrecognition of parasites as foreign invaders, was the basic idea of how molecular mimicry could work to favor parasite survival in immunocompetent hosts (SPRENT 1962; DINEEN 1963; DAMIAN 1964). The fundamental question is how eclipsed epitopes work, and this is related if not identical to the question of self-tolerance induction. At least for peptide epitopes, understanding is rapidly taking shape. A significant finding is that histocompatibility proteins function as cell receptors by binding semi-homologous immunogenic peptides for antigen presentation (GUILLET et al. 1987). A peptide with too much sequence homology to the antigen-presenting cell's binding receptor beta chain failed to stimulate T cells, leading these authors to postulate the existence of "holes in the T-cell repertoire." This immediately suggests a mechanism whereby parasite epitopes could become eclipsed, i.e., fail to stimulate an immune response (DAMIAN 1987b). It also suggests that if the parasite's peptide epitopes are not perfectly homologous to the appropriate peptide receptor-restriction element, then stimulation ensues, and they become potentially autoimmunogenic epitopes.

Efferent limb evasion through molecular mimicry would involve mechanisms to bypass a stimulated immune response. Again, some ideas on how this could occur were already presented. First, if autoimmunogenic epitopes are involved, they could participate in the genesis of immune response suppression to control chronic pathogenesis, with concomitant escape from a down-regulated immune response, as suggested by MITCHISON and OLIVEIRA (1986). Also, receptor mimicry could enter at this stage by enabling antigen masquerade (SMITHERS et al. 1968) or some other, more aggressive mode of evasion (DAMIAN 1984; GROSSMAN et al. 1986; KEMP et al. 1986) to take place.

8 Conclusions

Molecular mimicry between protozoan and helminthic endoparasites and their hosts is a common phenomenon. Its consequences to the host-parasite relationship are thought to be primarily fourfold: (1) induction of autoimmunity, (2) use as an immunological escape mechanism by the parasite, (3) use to gain the intracellular environment, and (4) development and maintenance of host antigenic polymorphisms as a host defensive mechanism. Much recent evidence for molecular mimicry as an escape mechanism comes from work on ligand and receptor mimicry by intracellular protozoan parasites. Through molecular mimicry, these parasites have an enhanced ability to invade cells, giving them a secure niche in the body where immune stimulation is at a minimum and where their metabolic requirements are met.

Extracellular helminth parasites may use molecular mimicry to avoid immune recognition, or if recognized, to bypass the stimulated immune response through host molecule masquerade or some other stratagem involving active interference with the host's immunoregulatory mechanism. The latter would also be subserved through receptor mimicry, enabling the parasite to cover itself in a specific way with host molecules such as immunoglobulins, complement components, or adhesive molecules such as FN. Hard evidence that any of these are really effective escape mechanisms in vivo remains to be gathered.

Molecular mimicry by parasites has probably contributed to the development and maintenance of host antigen gene polymorphisms, including immunosuppression gene polymorphisms, in important ways.

Acknowledgements. I am grateful to Drs. Louis H. Miller, Daniel J. Murfin, Kwame Nyame, and Rick L. Tarleton for helpful discussions. Much of the original work from my laboratory, and the writing of this review, was supported by grant number Al-18906 and previous grants from the U.S. National Institutes of Health.

References

Anders RF, Coppel RL, Brown GV, Kemp DJ (1988) Antigens with repeated amino acid sequences from the asexual blood stages of *Plasmodium falciparum*. Prog Allergy 41: 148–172

Ardeshir F, Flint JE, Richman SJ, Reese RT (1987) A 75 kd merozoite surface protein of *Plasmodium falciparum* which is related to the 70 kd heat-shock proteins. EMBO J 6: 493–499

Bayne CJ, Boswell CA, Yui MA (1987) Widespread antigenic cross-reactivity between plasma proteins of a gastropod, and its trematode parasite. Dev Comp Immunol 11: 321–329

Benacerraf B, McDevitt HO (1972) Histocompatibility-linked immune response genes. Science 175: 273–279

Ben-Ismail R, Carme B, Mogahed A, Niel G, Gentilini M (1982a) Antigen sharing between *Fasciola hepatica* and human erythrocytes. Tropenmed Parasitol 33: 11–14

Ben-Ismail R, Mulet-Clamagirand C, Carme B, Gentilini M (1982b) Biosynthesis of A, H, and Lewis blood group determinants in *Fasciola hepatica*. J Parasitol 68: 402–407

Bianco AE, Favaloro JM, Burkot TR, Culvenor JG, Crewther PE, Brown GV, Anders RF, Coppel RL, Kemp DJ (1986) A repetitive antigen of *Plasmodium faciparum* that is homologous to heat shock protein 70 of *Drosophila melanogaster*. Proc Natl Acad Sci USA 83: 8713–8717

Blackwell JM (1985) Receptors and recognition mechanisms of *Leishmania* species. Trans Roy Soc Trop Med Hyg 79: 606–612

Blackwell JM, Freeman J, Bradley D (1980) Influence of H-2 complex on acquired resistance to *Leishmania donovani* infection in mice. Nature 283: 72–74

Blackwell JM, Roberts MB (1987) Immunomodulation of murine visceral leishmaniasis by administration of monoclonal anti-Ia antibodies: differential effects of anti-l-A vs. anti-l-E antibodies. Eur J Immunol 17: 1669–1672

Bloom BR (1979) Games parasites play. Nature 279: 21–26

Capron A, Biguet J, Rose F, Vernes A (1965) Les antigenes de *Schistosoma mansoni*: II. Etude immunoelectrophoretique comparée de divers stades larvaires et des adultes des deux sexes. Aspects immunologiques des relation hote-parasite de la cercaire et de l'adulte de *S. mansoni*. Ann Inst Pasteur, Paris 109: 798–810

Capron A, Biguet J, Vernes A, Afchain D (1968) Structure antigenique des helminthes. Aspects immunologiques des relations hote-parasite. Pathol Biol 16: 121–138

Chang K-P, Fong D (1983) Cell biology of host-parasite membrane interactions in leishmaniasis. Ciba Found Symp 99: 113–137

Chappell TG, Welch WJ, Schlossman DM, Palter KB, Schlesinger MJ, Rothman JE (1986) Uncoating ATPase is a member of the 70 kilodalton family of stress proteins. Cell 45: 3–13

Cheung A, Leban J, Shaw AR, Merkli B, Stocker J, Chizzolini C, Sander C, Perrin LH (1986) Immunization with synthetic peptides of a *Plasmodium falciparum* surface antigen induces antimerozite antibodies. Proc Natl Acad Sci USA 83: 8328–8332

Chirico WJ, Waters MG, Blobel G (1988) 70K heat shock related proteins stimulate protein translocation into microsomes. Nature 332: 805–810

Cohen C, Reinhardt B, Castellani L, Norton P, Stirewalt M (1982) Schistosome spines are "crystals" of actin. J Cell Biol 61: 987–988

Dame JB, Williams JL, McCutchan TF, Weber JL, Wirtz RA, Hockmeyer WT, Maloy WL, Haynes JD, Schneider I, Roberts D, Sanders GS, Reddy EP, Diggs CL, Miller LH (1984) Structure of the gene encoding the immunodominant surface antigen on the sporozoite of the human malaria parasite *Plasmodium falciparum*. Science 225: 593–599

Damian RT (1962) A theory of immunoselection for eclipsed antigens of parasites and its implications for the problem of antigenic polymorphism in man. J Parasitol 48 (2, sec 2): 16 (abstract)

Damian RT (1964) Molecular mimicry: antigen sharing by parasite and host and its consequences. Am Naturalist 98: 129–149

Damian RT (1967) Common antigens between *Schistosoma mansoni* and the laboratory mouse. J Parasitol 53: 60–64

Damian RT (1979) Molecular mimicry in biological adaptation. In: Nickol BB (ed) Host-parasite interfaces: at population, individual and molecular levels. Academic, New York, pp 103–126

Damian RT (1984) Immunity in schistosomiasis: a holistic view. In: Marchalonis JJ (ed) Immunobiology of parasites and parasitic infections. Contemp Top Immunobiol 12: 359–420

Damian RT (1987a) The exploitation of host immune responses by parasites. J Parasitol 73: 3–13
Damian RT (1987b) Molecular mimicry revisited. Parasitol Today 3: 263–266
Davis AH, Blanton R, Klich P (1985) Stage and sex specific differences in actin gene expression in *Schistosoma mansoni*. Mol Biochem Parasitol 17: 289–298
Decker-Jackson JE, Honigberg BM (1978) Glycoproteins released by *Leishmania donovani*: immunologic relationships with host and bacterial antigens and preliminary biochemical analysis. J Protozool 25: 514–525
Deshaies RJ, Koch BD, Werner-Washburne M, Craig EA, Schekman R (1988) A subfamily of stress proteins facilitates translocation of secretory and mitochondrial precursor polypeptides. Nature 332: 800–805
DeVay JE, Adler HE (1976) Antigens common to hosts and parasites. Ann Rev Microbiol 30: 147–168
Dineen JK (1963) Antigenic relationship between host and parasite. Nature 197: 471–472
Dissous C, Grzych JM, Capron A (1986) *Schistosoma mansoni* shares a protective oligosaccharide epitope with freshwater and marine snails. Nature 323: 443–445
Dixit VM, Hennessy SW, Grant GA, Rotwein P, Frazier WA (1986) Characterization of a cDNA encoding the heparin and collagen binding domains of human thromboplastin. Proc Natl Acad Sci USA 83: 5449–5453
Doenhoff MJ, Hassounah OA, Lucas SB (1985) Does the immunopathology induced by schistosome eggs potentiate parasite survival? Immunol Today 6: 203–206
Dragon EA, Sias SR, Kato EA, Gabo JD (1987) The genome of *Trypanosoma cruzi* contains a constitutively expressed, tandemly arranged, multicopy gene homologous to a major heat shock protein. Mol Cell Biol 7: 1271–1275
Dyrberg T, Oldstone MBA (1986) Peptides as probes to study molecular mimicry and virus-induced autoimmunity. Curr Top Microbiol Immunol 134: 25–37
Ebringer A, Deacon NJ, Young CR (1976) Codominant inheritance in immunogenetic (IR-gene) systems. J Immunogenetics 3: 401–409
Fedotova TI, Gromova BB-O (1974) Parasitism in phytopathogenic organisms (in Russian). VNIITEISKH (All-Union Scientific Research Institute for Information and Technical and Economic Research in Agriculture), Moscow
Findlay DR, Newmeyer DD, Price TM, Forbes DJ (1987) Inhibition of in vitro nuclear transport by a lectin that binds to nuclear pores. J Cell Biol 104: 189–200
Frazier WA (1987) Thrombospondin: a modular adhesive glycoprotein of platelets and nucleated cells. J Cell Biol 105: 625–632
Fujinami RS, Oldstone MBA (1985) Amino acid homology between the encephalitogenic site of myelin basic protein and virus: mechanism for autoimmunity. Science 230: 1043–1045
Gasser DL (1969) Genetic control of the immune response in mice. I. Segregation data and localization to the fifth linkage group of a gene affecting antibody production. J Immunol 103: 66–70
Glass DJ, Polvere RI, Van der Ploeg LHT (1986) Conserved sequences and transcription of the hsp 70 gene family in *Trypanosoma brucei*. Mol Cell Biol 6: 4657–4666
Goundis D, Reid KBM (1988) Properdin, the terminal complement components, thrombospondin and the circumsporozoite protein of malaria parasites contain similar sequence motifs. Nature 335: 82–85
Greenblatt CL (1983) Molecular mimicry and the carbohydrate language of parasitism. ASM News 49: 488–493
Greenblatt CL, Kark JD, Schnur LF, Slutzky GM (1981) Do leishmania serotypes mimic human blood group antigens? Lancet ii: 505–506 (letter)
Greenblatt CL, Meline D, Slutzky GM, Schnur LF, Levine C (1984) Surface reaction of *Leishmania*. III. *Ulex europaeus* II lectin affinity for excreted factor (EF) serotype A strains. Ann Trop Med Parasitol 78: 99–107
Grossman Z, Greenblatt CL, Cohen IR (1986) Parasite immunology and lymphocyte population dynamics. J Theor Biol 121: 129–139
Guillet J-G, Lai M-Z, Briner TJ, Buus S, Sette A, Grey HM, Smith JA, Gefter ML (1987) Immunological self, nonself discrimination. Science 235: 865–870
Haas IG, Wabl M (1983) Immunoglobulin heavy chain binding protein. Nature 306: 387–389
Hadley TJ, Klotz FW, Miller LH (1986) Invasion of erythrocytes by malaria parasites: a cellular and molecular overview. Ann Rev Microbiol 40: 451–477

Haldane JBS (1949) Disease and evolution. In: Symposium sui fattori ecologici e genetici della speciazione negli animali, Pallanza, Italy, 1948. La Ricerca Scientifica 19 [Suppl]: 68–76

Handman E, Goding JW (1985) The *Leishmania* receptor for macrophages is a lipid-containing glycoconjugate. EMBO J 4: 329–336

Hedstrom R, Culpepper J, Harrison RA, Agabian N, Newport G (1987) A major immunogen in *Schistosoma mansoni* infections is homologous to the heat-shock protein Hsp70. J Exp Med 165: 1430–1435

Hirayama K, Matsushita S, Kikuchi I, Iuchi M, Ohta N, Sasazuki T (1987) HLA-DQ is epistatic to HLA-DR in controlling the immune response to schistosomal antigen in humans. Nature 327: 426–430

Holmquist G, Udomsangpetch R, Berzins K, Wigzell H, Perlmann P (1988) *Plasmodium chabaudi* antigen Pch105, *Plasmodium falciparum* antigen Pf155, and erythrocyte band 3 share cross-reactive epitopes. Infect Immun 56: 1545–1550

Holt GD, Hart GW (1986) The subcellular distribution of terminal N-acetylglucosamine moieties. J Biol Chem 261: 8049–8057

Jack RM, Ward PA (1980) The role *in vivo* of C3 and the C3b receptor in babesial infection in the rat. J Immunol 124: 1574–1578

Jahnke U, Fischer EH, Alvord EC Jr (1985) Sequence homology between certain viral proteins and proteins related to encephalomyelitis and neuritis. Science 229: 282–284

Kaplan MH, Meyeserian M (1962) An immunological cross-reaction between group-A streptococcal cells and human heart tissue. Lancet i: 706–710

Kaslow DC, Quakyi IA, Syin C, Raum MG, Keister DB, Coligan JE, McCutchan TF, Miller LH (1988) A vaccine candidate from the sexual stage of human malaria that contains EGF-like domains. Nature 333: 74–76

Kemp WM, Tarleton RL, McGuiness TB, Rasmussen KR, Devine DV (1986) Biology of tegument associated IgG-Fc and C3 receptors in *Schistosoma mansoni*. J Chem Ecol 12: 1833–1841

Koyasu S, Nishida E, Kadowaki T, Matsuzaki F, Iida K, Harada F, Kasuga M, Sakai H, Yahara I (1986) Two mammalian heat shock proteins, HSP90 and HSP100, are actin-binding proteins. Proc Natl Acad Sci USA 83: 8054–8058

Lawler J, Hynes RO (1986) The structure of human thrombospondin, an adhesive glycoprotein with multiple calcium-binding sites and homologies with several different proteins. J Cell Biol 103: 1635–1648

Lernmark Å, Dyrberg T, Terenius L, Hokfelt B (1988) (eds) Molecular mimicry in health and disease. Nordisk Insulin Symp No 2-ICS 823. Elsevier, Amsterdam

Lindquist S (1986) The heat-shock response. Ann Rev Biochem 55: 1151–1191

Linthicum DS, Bolger MB (1985) Using molecular mimicry to produce anti-receptor antibodies. Bioassays 3: 213–217

Linthicum DS, Nadir RF (1988) (eds) Anti-idiotypes, receptors, and molecular mimicry. Springer, Berlin Heidelberg New York

Marchalonis JJ, Morris PJ, Harris AW (1974) Speculations on the functions of immune response genes. J Immunogenetics 1: 63–77

Markowitz AS, Armstrong JS, Kushner DS (1960) Immunological relationship between the rat glomerulus and nephritogenic streptococci. Nature 187: 1095–1097

Matsumoto Y, Perry G, Levine RJC, Blanton R, Mahmoud AAF, Aikawa M (1988) Paramyosin and actin in schistosomal teguments. Nature 333: 76–78

McCormick SL, Damian RT (1987) Haptenation of adult *Schistosoma mansoni* and assessment of humorally mediated damage in vitro. J Parasitol 73: 130–143

McGuiness TB, Kemp WM (1981) *Schistosoma mansoni*: evidence for a complement dependent receptor on adult male parasites. Exp Parasitol 51: 236–242

McLaughlin GL, Benedik MJ, Campbell GH (1987) Repeated immunogenic amino acid sequences of *Plasmodium* species share sequence homologies with proteins from humans and human viruses. Am J Trop Med Hyg 37: 258–262

Mitchison NA, Oliveira DBG (1986) Chronic infection as a major force in the evolution of the suppressor T-cell system. Parasitol Today 2: 312–313

Mosser DM, Edelson PJ (1985) The mouse macrophage receptor for C3bi (CR3) is a major mechanism in the phagocytosis of *Leishmania* promastigotes. J Immunol 135: 2785–2789

Munro S, Pelham HRB (1986) An Hsp70-like protein in the ER: identity with the 78 kd glucose-regulated protein and immunoglobulin heavy chain binding protein. Cell 46: 291–300

Nene V, Dunne DW, Johnson KS, Taylor DW, Cordingly JS (1986) Sequence and expression of a major egg antigen from *Schistosoma mansoni*. Homologies to heat shock proteins and alpha-crystallins. Mol Biochem Parasitol 21: 179–188

Nyame K, Cummings RD, Damian RT (1987) *Schistosoma mansoni* synthesizes glycoproteins containing terminal *O*-linked *N*-acetylglucosamine residues. J Biol Chem 262: 7990–7995

Nyame K, Cummings RD, Damian RT (1988a) Characterization of the high mannose asparagine-linked oligosaccharides synthesized by *Schistosoma mansoni* adult male worms. Mol Biochem Parasitol 28: 265–274

Nyame K, Cummings RD, Damian RT (1988b) Characterization of the *N*- and *O*-linked oligosaccharides in glycoproteins synthesized by *Schistosoma mansoni* schistosomula. J Parasitol 74: 562–572

Nyame K, Smith DF, Cummings RD, Damian RT (1989) Complex-type asparagine-linked oligosaccharides synthesized by *Schistosoma mansoni* adult males contain terminal β-linked *N*-acetylgalactosamine. J Biol Chem 264: 3235–3243

Ofek I, Sharon N (1988) Lectinophagocytosis: a molecular mechanism of recognition between cell surface sugars and lectins in the phagocytosis of bacteria. Infect Immun 56: 539–547

Oldstone MBA (1987) Molecular mimicry and autoimmune disease. Cell 50: 819–820

Ouaissi MA (1988) Role of the RGD sequence in parasite adhesion to host cells. Parasitol Today 4: 169–173

Ouaissi MA, Cornette J, Capron A (1984) Occurrence of fibronectin antigenic determinants on *Schistosoma mansoni* lung schistosomula and adult worms. Parasitology 88: 85–96

Ouaissi MA, Cornette J, Afchain D, Capron A, Gras-Masse H, Tartar A (1986) *Trypanosoma cruzi* infection inhibited by peptides modeled from a fibronectin cell attachment domain. Science 234: 603–607

Polla BS (1988) A role for heat shock proteins in inflammation? Immunol Today 9: 134–136

Ribaux CL, Magloire H, Joffre A, Morrier JJ (1983) Immunohistochemical localization of fibronectin-like protein on the cell surface of the oral flagellate *Trichomonas tenax*. J Biol Buccale 11: 41–51

Roberts DD, Sherwood JA, Spitalnik SL, Panton LJ, Howard RJ, Dixit VM, Frazier WA, Miller LH, Ginsburg V (1985) Thrombospondin binds falciparum malaria parasitized erythrocytes and may mediate cytoadherence. Nature 318: 64–66

Robson KJH, Hall JRS, Jennings MW, Harris TJR, Marsh K, Newbold CI, Tate VE, Weatherall DJ (1988) A highly conserved amino-acid sequence in thrombospondin, properdin and in proteins from sporozoites and blood stages of a human malaria parasite. Nature 335: 79–82

Rowley D, Jenkin CR (1962) Antigenic cross-reaction between host and parasite as a possible cause of pathogenicity. Nature 193: 151–154

Rumjanek FD, Smithers SR (1978) Mannosyl transferase activity in homogenates of adult *Schistosoma mansoni*. Parasitology 77: 75–86

Rumjanek FD, Broomfield KE, Smithers SR (1978) *Schistosoma mansoni*: glycosyl transferase activity and the carbohydrate composition of the tegument. Exp Parasitol 47: 24–35

Ruoslahti E, Pierschbacher MD (1987) New perspectives in cell adhesion: RGD and integrins. Science 238: 491–497

Russell DG, Wilhelm H (1986) The involvement of the major surface glycoprotein (gp63) of *Leishmania* promastigotes in attachment to macrophages. J Immunol 136: 2613–2620

Sacks DL, Kirchhoff LV, Hieny S, Sher A (1985) Molecular mimicry of a carbohydrate epitope on a major surface glycoprotein of *Trypanosoma cruzi* by using anti-idiotypic antibodies. J Immunol 135: 4155–4159

Santoro F, Ouaissi MA, Capron A (1979) Receptors for complement (C1q and C3b) on the immature forms of *Schistosoma mansoni*. ICRS Med Sci 7: 576

Sasazuki T, Matsushita S, Muto M, Ohta N (1983) HLA-linked genes controlling the immune response and disease susceptibility. Immunol Rev 70: 51–75

Schad GA (1966) Immunity, competition, and natural regulation of helminth populations. Am Naturalist 100: 359–364

Selkirk ME, Rutherford PJ, Denham DA, Partono F, Maizels RM (1987) Cloned antigen genes of *Brugia* filarial parasites. Biochem Soc Symp 53: 91–102

Smithers SR, Terry RJ, Hockley DJ (1968) Do adult schistosomes masquerade as their hosts? Trans Roy Soc Trop Med Hyg 62: 466–467

Snell GD (1968) The H-2 locus of the mouse: observations and speculations concerning its comparative genetics and its polymorphism. Folia Biol (Prague) 14: 335–358

Sprent JFA (1962) Parasitism, immunity and evolution. In: Leeper GW (ed) The evolution of living organisms. Symposium of the Royal Society of Melbourne, 1959. Melbourne University Press

Srinivasappa J, Saegusa J, Prabhakar BS, Gentry MK, Buchmeier MJ, Wiktor TJ, Koprowski H, Oldstone MBA, Notkins AL (1986) Molecular mimicry: frequency of reactivity of monoclonal antiviral antibodies with normal tissues. J Virol 57: 397–401

Stowe MK (1988) Chemical mimicry. In: Spencer KC (ed) The chemical mediation of coevolution. Pergamon, New York

Tarleton RL, Kemp WM (1981) Demonstration of IgG-Fc and C3 receptors on adult *Schistosoma mansoni*. J Immunol 126: 379–384

Torpier G, Capron A, Ouaissi MA (1979) Receptor for IgG (Fc) and human β_2-microglobulin on *S. mansoni* schistosomula. Nature 278: 447–449

Torres C-R, Hart GW (1984) Topography and polypeptide distribution of terminal *N*-acetylglucosamine residues on the surfaces of intact lymphocytes. Evidence for *O*-linked GlcNAc. J Biol Chem 259: 3308–3317

Turco SJ (1988) The lipophosphoglycan of *Leishmania*. Parasitol Today 4: 255–257

Van der Ploeg LHT, Giannini SH, Cantor CR (1985) Heat shock genes: regulatory role for differentiation in parasitic protozoa. Science 228: 1443–1446

Vane-Wright RI (1976) A unified classification of mimetic resemblances. Biol J Linn Soc 8: 25–56

Wassom DL, Krco CJ, David CS (1987) I-E expression and susceptibility to parasite infection. Immunol Today 8: 39–43

Wickler W (1968) Mimicry in plants and animals. McGraw-Hill, New York (English translation)

Wright MD, Tiu WU, Wood SM, Walker JC, Garcia EG, Mitchell GF (1988) *Schistosoma mansoni* and *S. japonicum* worm numbers in 129/J mice of two types and dominance of susceptibility in F_1 hybrids. J Parasitol 74: 618–622

Wyler DJ, Sypek JP, McDonald JA (1985) In vitro parasite-monocyte interactions in human leishmaniasis: possible role of fibronectin in parasite attachment. Infect Immun 49: 305–311

Yang Y-F, Tan-ariya P, Sharma YD, Kilejian A (1987) The primary structure of a *Plasmodium falciparum* polypeptide related to heat shock proteins. Mol Biochem Parasitol 26: 61–68

Yoshino TP, Boswell CA (1986) Antigen sharing between larval trematodes and their snail hosts: how real a phenomenon in immune evasion? Symp Zool Soc London 56: 221–238

Young D, Lathigra R, Hendrix R, Sweetser D, Young RA (1988) Stress proteins are immune targets in leprosy and tuberculosis. Proc Natl Acad Sci USA 85: 4267–4270

Yuckenberg PD, Poupin F, Mansour TE (1987) *Schistosoma mansoni*: protein composition and synthesis during early development; evidence for early synthesis of heat shock proteins. Exp Parasitol 63: 301–311

Molecular Mimicry and Diabetes

T. Dyrberg

1 Introduction   117
2 Insulin-Dependent Diabetes Mellitus: Etiology and Pathogenesis   117
3 Autoimmune Insulin-Resistant Diabetes — Identification and Implications of Shared Peptide Sequences Between the Insulin Receptor and Virus Proteins   119
4 Lessons from Other Diseases: Myasthenia Gravis and Autoimmune Thyroditis   122
5 Concluding Remarks   123
References   123

1 Introduction

Insulin-dependent diabetes (IDDM) often develops early in life and is associated with lifelong insulin treatment, a reduced life span, and serious long-term complications (Borch-Johnsen et al. 1986). Other forms of diabetes exist (National Diabetes Data Group 1979), e.g., noninsulin-dependent diabetes, maturity onset diabetes in the young (MODY), and insulin-resistant diabetes associated with a decreased function of the insulin receptor. In this review we discuss the findings of molecular mimicry between virus proteins and autoantigens in IDDM and insulin-resistant diabetes associated with antibodies to the insulin receptor and the implications for its etiology and pathogenesis. Autoimmunity has in both diseases been shown to be important in the pathogenesis, but little is known about the initiating factors. Observations and analysis of homologies between viral antigens and autoantigens involved in the disease offer an unique way of probing for possible etiologic agents.

2 Insulin-Dependent Diabetes Mellitus: Etiology and Pathogenesis

The pathogenesis of IDDM involves an autoimmune-mediated destruction of the insulin-producing β cells in the islets of Langerhans (reviewed in Rossini et al. 1985). Evidence for an autoimmune pathogenesis comes from a large number of observations

---

Hagedorn Research Laboratory, Niels Steensensvej 6, DK-2820 Gentofte, Denmark

of which only a few will be mentioned here. At the clinical onset of IDDM, the islets are massively infiltrated by mononuclear cells, but the immune response is specific, since only the β cells and not the other endocrine islet cells are affected. Alternatively, the β cells may be particularly susceptible to substances produced by the activated immunocompetent cells like interleukin-1 (MANDRUP-POULSEN et al. 1986). In any case, autoantibodies are commonly found at onset of the disease, either reacting specifically with the surface of β cells, presumably recognizing a M_r 64000 plasma membrane protein, or binding to a cytoplasmic antigen common to all endocrine islet cells. Susceptibility for development of IDDM is strongly associated with certain molecules coded for by HLA immune response genes from the HLA DR and, in particular, the HLA DQ locus.

Studies of IDDM in homozygous twins show a concordance rate of less than 50% (BARNETT et al. 1981) indicating that exogenous factors play a role in initiating the disease, and viruses in particular have been implicated (TONIOLO and ONODERA 1984; LEITER and WILSON 1988). Many studies have shown correlations between high levels of virus antibodies, notably to mumps, coxsackie B4, and rubella virus, and the development of IDDM. Direct association between virus and IDDM has been shown in only a few cases, e.g., infection with a coxsackie B4 virus followed by acute onset of IDDM (YOON et al. 1979). Current knowledge points, however, to the fact that the clinical onset of IDDM is preceded by a longer latent period characterized by gradually diminishing β cell function and the presence of islet cell antibodies (ROSSINI et al. 1985). Exposure to an etiologic agent may thus have taken place months or years before the onset of IDDM, at which time the agent itself may have disappeared. In this context, it is interesting that congenital infection with rubella virus is associated with an increased incidence of IDDM, which develops 15–20 years after infection and is indistinguishable from spontaneous IDDM (GINSBERG-FELLNER et al. 1985).

Viruses may initiate cell destructive processes by several mechanisms (McCHESNEY and OLDSTONE 1987): (a) By a primary infection, acute or persistent, of β cells causing a decrease in cell function or cell destruction followed by a secondary immune reaction to the β cells' constituents; (b) inducing HLA class II antigen expression on β cells, thereby, they could present antigen and initiate an autoimmune response; (c) by infecting selcted subsets of lymphocytes leading to defects in cell regulation or effector mechanisms; (d) by inducing the expression of new surface antigens or altering existing ones or by the presence of homologous antigenic epitopes on virus proteins and β cell autoantigens, i.e., molecular mimicry (OLDSTONE 1987). Immune recognition of such components could result in a break down of immune self-tolerance and lead to β cell destruction.

The target antigen for the β cell destructive autoimmune response in IDDM has not yet been identified. One obvious candidate for such an antigen is the β cell-specific product, insulin, and autoantibodies to insulin have in fact been demonstrated in IDDM before treatment with exogenous insulin is initiated (PALMER et al. 1983). Intracisternal type A retroviral particles or their gene products have been implicated in the pathogenesis of diabetes in mice (LEITER and WILSON 1988), since they are found in the β cells of certain diabetes-prone mouse strains and the number of viral particles increases with the development of hyperglycemia. Insulin-specific antibodies have also been demonstrated in diabetes-prone mice. Recently it was shown in NOD mice, a model for autoimmune IDDM, and in C57B/KsJ db/db mice, a model for insulin-resistant diabetes, that insulin-specific antibodies cross-react with the retroviral antigen p73 of intracisternal type A retroviral particles (SERREZE et al. 1988). Absorption studies indicate the presence of a shared antigenic epitope on insulin and the retroviral p73 antigen and its homologous cellular counterpart, an IgE-binding factor. No pathogenic role is known for these cross-reactive antibodies, and they could simply appear secondary to β cell damage, but their presence is intriguing and deserves further study. Antibodies to retroviral *gag* gene products can be demonstrated in many patients with autoimmune connective tissue diseases. It is, therefore, interesting that a common amino acid sequence has been identified between the p30^{gag} of mammalian type C retroviruses and an autoantigen found in individuals with systemic lupus erythematosus and mixed connective tissue disease, the 70K protein component of small ribonucleic proteins (QUERY and KEENE 1987). Endogenous or acquired retroviruses and their gene products may prove to be an important source of immunogenic stimulus for autoimmune responses.

Autoantibodies to the insulin receptor have also been demonstrated in IDDM patients (MARON et al. 1983). The amino acid sequence of the receptor has been determined, which allows a detailed analysis of the autoimmune response to the receptor, including a search for shared epitopes between the receptor and virus proteins. In IDDM, however, the insulin receptor is not likely to be an important autoantigen and the antibody response to the receptor may occur as an anti-idiotype reaction to insulin antibodies (SHOELSON et al. 1986).

One candidate for an important autoantigen in IDDM is the M_r 64000 islet cell protein. Antibodies to this component are found with a high prevalence in IDDM patients, also prior to the onset of the disease (BÆKKESKOV et al. 1987). Once the molecular structure of this protein becomes available, it will be possible to dissect the autoimmune response and analyze what pathogenic role it plays in the development of IDDM. Further, determination of the amino acid sequence of this protein will make it possible to look directly for homologous sequences between it and virus proteins, which may give a molecular basis for the association with those viruses particularly implicated in the etiology of IDDM, i.e., rubella, coxsackie B4, and mumps viruses.

3 Autoimmune Insulin-Resistant Diabetes — Identification and Implications of Shared Peptide Sequences Between the Insulin Receptor and Virus Proteins

Some rare cases of diabetes are characterized by marked insulin resistance without β cell destruction. This syndrome has been subdivided into two groups, A and B (KAHN et al. 1976). The pathogenesis of the B type has convincingly been shown to involve specific autoantibodies binding to the insulin receptor and interfering with its function (KAHN and HARRISON 1981). This type of insulin resistance is further associated with other autoimmune manifestations and acanthosis nigricans, a skin disease of unknown origin. Thus, although this is a rare form of diabetes, it is

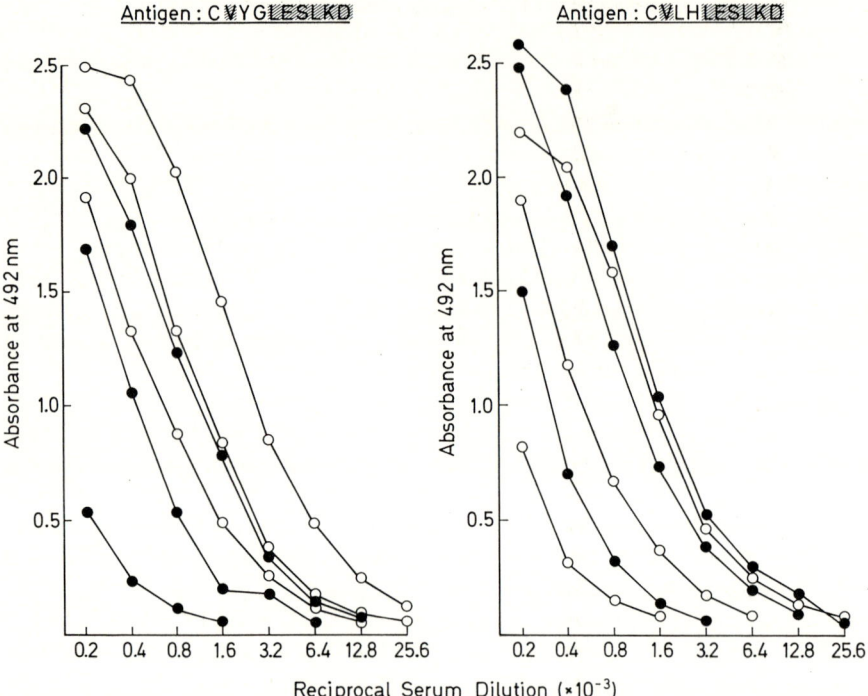

Fig. 1. Rats were immunized with peptides representing either the α-chain of the human insulin receptor (HINSR residues *66–74*) or the E2 protein of *papillomavirus* type 1a, residues 76–84. The peptides were synthesized with an *N*-terminal cysteine residue through which the peptides were coupled to keyhole limpet hemocyanin as a carrier protein. Binding of immune sera to the peptide used for immunization and to the homologous peptide was tested by ELISA using uncoupled peptides as antigen. Preimmune sera induced binding below 0.1 OD units. In the *left panel* is shown binding to the insulin receptor peptide and in the *right panel*, binding to the papillomavirus peptide

important because the autoimmune character of the pathogenesis is well-elucidated, and the primary structure of the autoantigen is known (ULLRICH et al. 1985). The initiating factor of this disease is, however, completely unknown. We have previously reported the presence of several significant amino acid homologies between the extra cytoplasmic region of the insulin receptor and viral proteins (DYRBERG and OLDSTONE 1986). In order to analyze further these homologies, we synthesized the respective amino acid sequences, immunized rats with the peptides coupled to carrier proteins, and tested the immune sera by ELISA for binding to the peptides used for immunization and for cross-reactivity with the homologous peptides. In Fig. 1 and 2 are shown two examples. In the first example, rats were immunized either with a peptide representing the insulin receptor, residues 66–74 (VYGLESLKD), or with a homologous peptide sharing seven of the nine residues, representing the E2 protein of papillomavirus type 1a residues 76–84 (VLHLESLKD). Antisera to the receptor

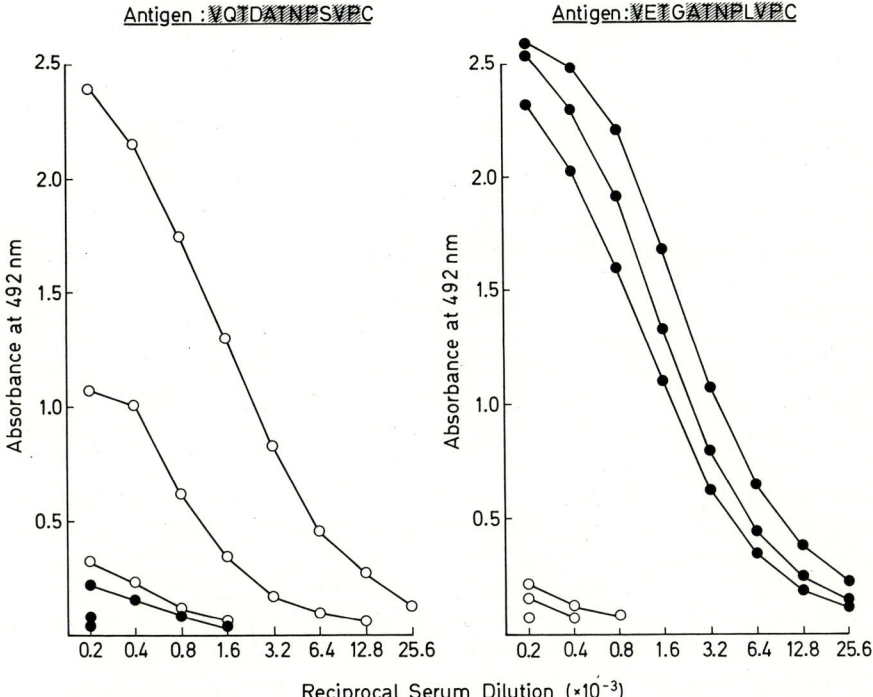

Fig. 2. Rats were immunized with peptides representing either the α-chain of the human insulin receptor (*HINSR* residues *587–597*) or the VP1 protein of *poliovirus*, residues 625–635. Peptides were synthesized with a C-terminal cysteine residue and coupled to keyhole limpet hemocyanin. Immune sera were tested for binding to uncoupled peptides by ELISA. In the *left panel* is shown binding to the insulin receptor peptide and in the *right panel*, to poliovirus peptide

peptide reacted as well with the immunizing receptor peptide as with the virus peptide, and similarly antisera to the papillomavirus peptide showed equal binding to both the virus and the receptor peptide (Fig. 1). In another example (Fig. 2), antisera from two of three rats immunized with a peptide representing the α-chain of the insulin receptor, residues 587–597 (VQTDATNPSVP), showed strong binding to the receptor peptide but little or no binding to a mimicking peptide representing the VP1 protein of poliovirus, residues 625–635 (VETGATNPLVP). All rats immunized with the poliovirus peptide reacted strongly in ELISA with the virus peptide but not with the homologous receptor peptide. These data suggest that factors other than the number of identical amino acids in an epitope determine whether an immune response is cross-reactive or not.

The homology between the insulin receptor and the papillomavirus is, at least, of theoretical interest because of the association between insulin-resistant diabetes and acanthosis nigricans, a pigmentous, wartlike skin disease of unknown etiology, and further that papillomavirus type la is one of the causative agents of some types of human warts. The significance of this homology will be analyzed along the lines previously described (DYRBERG and OLDSTONE 1986), including analysis of the peptide antibodies reactivity and cross-reactivity to the native molecules and the antibodies' ability to interfere with insulin binding to the insulin receptor. More than one region of the insulin receptor may, however, be involved in the generation of the autoimmune response in this type of diabetes, since it has been shown that autoantibodies from different patients bind to different parts of the receptor (FLIER et al. 1977).

4 Lessons from Other Diseases: Myastenia Gravis and Autoimmune Thyroiditis

A rational approach to dissect the mechanisms and investigate the initiating factors of a cell or organ-specific autoimmune disease requires knowledge of the molecular structure of the target antigen(s). Specific autoantigens have been identified in a number of putative autoimmune, organ-specific diseases (BÄRMEIER et al. 1988). However, in only a few cases have the antigens been isolated, their primary structure characterized, and a definitive correlation between the autoimmune response to the antigen and the ensuing disease established, e.g., in myasthenia gravis, which is caused by a humoral autoimmune response to the acetylcholine receptor (LINDSTROM et al. 1988). The etiology of mysthenia gravis is not known, but a virus has been suspected as a causative agent. To study whether a shared epitope between a virus protein and the acetylcholine receptor could be an initiating factor in myasthenia, we searched the NBRF protein data base (GEORGE et al. 1986) for homologies between the immunodominant α-chain of the receptor and virus proteins. A hydrophilic region of the α-chain showed a homology to an amino acid sequence of a glycoprotein from herpes simplex virus. Sera from myasthenic patients contained autoantibodies that bound to a synthetic peptide representing this particular sequence of the receptor, and such antibodies, affinity purified on the receptor peptide, specifically bound to herpes simplex virus-infected cells (SCHWIMMBECK et al. 1989). In another example, a six-amino acid sequence was shared between the α-chain of the acetylcholine receptor and an exposed epitope of a poliovirus coat protein. Antibodies to a synthetic peptide representing the homologous sequence were shown to bind both poliovirus and human acetylcholine receptor in solid phase ELISA (DYRBERG and OLDSTONE 1987). The kind of data cited above do not prove an etiologic role for viruses in these diseases, but they provide a basis for testing whether a given virus may be an initiating factor in an autoimmune response.

With the current advances in molecular biology, more and more autoantigens become available for similar analysis. In patients with autoimmune thyroid disease, autoantibodies to the thyroid microsomal antigen can often be detected, and it has been shown that this autoantigen is probably identical to thyroid peroxidase (PORTMANN et al. 1988). Recently, the amino acid sequence of thyroid peroxidase was obtained by cloning and sequencing (KIMURA et al. 1987). Comparison with the NBRF protein data base demonstrated a number of mimicking epitopes shared by the peroxidase and several human virus proteins (Table 1). Analysis of the cross-reactivity induced by these shared epitopes may help to find an etiological factor in some forms of autoimmune thyroid disease.

Table 1. Examples of shared amino acid sequences between the human thyroid peroxidase (thyroid microsomal antigen) and virus proteins

Protein	Residues	Amino acid sequence
Thyroid peroxidase Adenovirus 2 and 5, late 100K protein	241–250 794–803	P A A G T A C L P F G A A G T A C S P T
Thyroid peroxidase Adenovirus 2, penton protein (III)	325–332 38– 45	L P F V P P R R P P F V P P R Y
Thyroid peroxidase Hepatitis B virus, DNA polymerase	424–431 92– 99	F Q Q Y V G P Y C Q Q Y V G P L
Thyroid peroxidase Adenovirus 2 and 5, DNA polymerase	521–528 132–139	L T E R L F V L R T E R L F V T
Thyroid peroxidase Papillomavirus, type 33, E2 protein	579–588 11– 20	V A D K I L D L Y K V Q E K I L D L Y E

5 Concluding Remarks

One mechanism whereby virus can induce autoimmunity is through molecular mimicry between viral proteins and host cell proteins. Identification and study of homologies between autoantigens involved in the disease, e.g., the insulin receptor, on one side and infectious agents on the other make it possible to search directly for etiological factors. Infectious agents initiating an autoimmune disorder by such a mechanism may, at an early stage, be cleared from the host and thus be absent at the clinical onset of disease. In diabetes mellitus caused by autoimmune destruction of β cells, the major target antigen(s) remains to be characterized at the molecular level. Results obtained from examining similar autoimmune diseases may, however, give important information on the role of molecular mimicry in the development of autoimmunity and establish principles for how, on one hand, to analyze randomly occurring homologies between unrelated proteins and, on the other, to study spontaneous or experimentally occurring cross-reactive antibodies. Studies along these lines may implicate viruses as etiological factors of such autoimmune disorders, allowing the question of the virus-disease relationship to be directly addressed.

References

Barnett AH, Eff C, Leslie RDG, Pyke DA (1981) Diabetes in identical twins: a study of 200 pairs. Diabetologia 20: 87–93

Borch-Johnsen K, Kreiner S, Deckert T (1986) Mortality of type (insulin-dependent) diabetes mellitus in Denmark: a study of relative mortality in 2,930 Danish Type 1 diabetic patients diagnosed from 1933 to 1972. Diabetologia 29: 767–772

Bækkeskov S, Landin M, Kristensen JK, Srikanta S, Bruining GJ, Mandrup-Poulsen T, de Beaufort C, Soeldner JS, Eisenbarth G, Lindgren F, Sundquist G, Lernmark Å (1987) Antibodies to a 64,000 M_r human islet cell antigen precede the clinical onset insulin-dependent diabetes. J Clin Invest 79: 926–934

Bärmeier H, Christie M, Herold B, Herold K, Lernmark Å (1989) The humoral anti-islet response II. Biochemical studies. In: Ginsberg-Fellner F, McEvoy R (eds) Autoimmunity and the pathogenesis of diabetes. Springer, Berlin Heidelberg New York (in press)

Dyrberg T, Oldstone MBA (1986) Peptides as probes to study molecular mimicry and virus-induced autoimmunity. Curr Topic Microbiol Immunol 130: 25–37

Dyrberg T, Oldstone MBA (1987) Molecular mimicry between viral proteins and cell antigens — implications for development of autoimmunity. Period Biol 89 [Suppl 1]: 48 (abstract)

Flier JS, Kahn CR, Jarrett DB, Roth J (1977) Autoantibodies to the insulin receptor. Effect on the insulin-receptor interaction in IM-9 lymphocytes. J Clin Invest 60: 784–794

George DG, Barker WC, Hunt LT (1986) The protein identification resource (PIR). Nucleic Acids Res 14: 11–15

Ginsberg-Fellner F, Witt ME, Fedun B, Taub F, Dobersen MJ, McEvoy RC, Cooper LZ, Notkins AL, Rubinstein P (1985) Diabetes mellitus and autoimmunity in patients with the congenital rubella syndrome. Rev Infect Dis 7, suppl. 1: S170–S176

Kahn CR, Harrison LC (1981) Insulin receptor autoantibodies. Carbohydrate metabolism and its disorders 3: 279–330

Kahn CR, Flier JS, Bar RS, Archer JA, Gorden P, Martin MM, Rot J (1976) The syndromes of insulin resistance and acanthosis nigricans. N Engl J Med 294: 739–745

Kimura S, Kotani T, McBride OW, Umeki K, Hirai K, Nakayama T, Ohtaki S (1987) Human thyroid peroxidase: complete cDNA and protein sequence, chromosome mapping, and identification of two alternately spliced mRNAs. Proc Natl Acad Sci 84: 5555–5559

Leiter EH, Wilson GL (1988) Viral interactions with pancreatic beta cells. In: Lefebvre P, Pipeleers D (eds) The pathology of the endocrine pancreas in diabetes. Springer, Berlin Heidelberg New York, pp. 85–105

Lindstrom J, Shelton D, Fujii Y (1988) Myasthenia gravis. Adv Immunol 42: 233–284

Mandrup-Poulsen T, Bendtzen K, Nerup J, Dinarello CA, Svenson M, Nielsen JH (1986) Affinity-purified interleukin-1 is cytotoxic to isolated islets of Langerhans. Diabetologia 29: 63–67

Maron R, Elias D, de Jongh BM, Brunining GJ, van Rood JJ, Shechter Y, Cohen IR (1983) Autoantibodies to the insulin receptor in juvenile onset, insulin-dependent diabetes. Nature 303: 817–818

McChesney MB, Oldstone MBA (1987) Viruses perturb lymphocyte functions: elected principles characterizing virus-induced immunosuppression. Ann Rev Immunol 5: 279–304

National Diabetes Data Group (1979) Classification and diagnosis of diabetes mellitus and other categories of glucose intolerance. Diabetes 28: 1039–1057

Oldstone MBA (1987) Molecular mimicry and autoimmune disease. Cell 50: 819–820

Palmer JP, Asplin CM, Clemons P, Lyen K, Tatpati O, Raghu PK, Paquette TL (1983) Insulin antibodies in insulin-dependent diabetics before insulin treatment. Science 222: 1337–1339

Portmann L, Fitch FW, Havran W, Hamada N, Franklin WA, DeGroot LJ (1988) Characterization of the thyroid microsomal antigen and its relationship to thyroid peroxidase, using monoclonal antibodies. J Clin Invest 81: 1217–1224

Quercy CC, Keene JD (1987) A human autoimmune protein associated with U1 RNA contains a region of homology that is crossreactive with retroviral $p30^{gag}$ antigen. Cell 57: 211–220

Rossini AA, Morders JP, Like AA (1985) Immunology of insulin-dependent diabetes mellitus. Ann Rev Immunol 3: 289–320

Rucheton M, Graafland H, Fanton H, Ursule L, Ferrier P, Larsen CJ (1985) Presence of circulating antibodies against gag-gene MuLV proteins in patients with autoimmune connective tissue disorders. Virology 144: 468–480

Schwimmbeck PL, Dyrberg T, Drachman D, Oldstone MBA (1989) Molecular mimicry and myasthenia gravis: uncovering a novel site of the acetylcholine receptor that has biologic activity and reacts immunochemically with herpes simplex virus. J Clin Invest (submitted for publication)

Serreze DV, Leiter EH, Kuff EL, Jardieu P, Ishizaka K (1988) Molecular mimicry between insulin and retroviral antigen p73. Development of cross-reactive autoantibodies in sera of NOD and C57BL/KsJ db/db mice. Diabetes 37: 351–358

Shoelson SE, Marshall S, Horikoshi H, Kolterman OG, Rubenstein AH, Olefsky JM (1986) Antiinsulin receptor antibodies in an insulin-dependent diabetic may arise as autoantiidiotypes. J Clin Endocrinol Metab 63: 56–61

Toniolo A, Onodera T (1984) Viruses and diabetes. In: Andreani D, DiMario U, Federlin KF, Heding LG (eds) Immunology in diabetes. Kimpton Medical, London, pp. 71–93

Ullrich A, Bell JR, Chen EY, Herrera R, Petruzzelli LM, Dull T, Gray A, Coussens L, Liao YC, Tsubokawa M, Mason A, Seeburg PH, Grunfeld C, Rosen OM, Ramachandran J (1985) Human insulin receptor and its relationship to the tyrosine kinase family of oncogenes. Nature 313: 756–561

Yoon JW, Austin M, Onodera T, Notkins AL (1979) Virus induced diabetes mellitus: isolation of a virus from the pancreas of a child with diabetic ketoacidosis. N Engl J Med 300: 1173–1179

Molecular Mimicry as a Mechanism for the Cause and as a Probe Uncovering Etiologic Agent(s) of Autoimmune Disease

M. B. A. OLDSTONE

For the purposes of our investigation, molecular mimicry is defined as similar structures shared by molecules from dissimilar genes or by their protein products. Either the molecules' linear amino acid sequences or their conformational fit may be shared, even though their origins are as separate as, for example, a virus and a normal host self determinant (Fig. 1). Because guanine cytosine (GC) sequences and introns designed to be spliced away may provide, respectively, false hybridization signals and nonsense homologies, focus on molecular mimicry is necessary at the protein level. Such homologies between proteins have been detected either by use of immunologic reagents, humoral or cellular, that cross-react with two presumably unrelated protein structures, or by computer searches to match proteins described in storage banks. Regardless of the methods used for identification, it is now abundantly clear that molecular mimicry between proteins encoded by numerous microbes and host "self" proteins is a common event. Such data is not only of interest in autoimmunity but also as a likely mechanism by which viral proteins are processed inside cells (Fig. 2; DALES et al. 1983). The cartoon of Fig. 2 attempts to bring together a number of observations including (a) the high incidence of cytoskeletal autoantibodies found in most microbial (viral) infections and (b) increasing observations that viruses

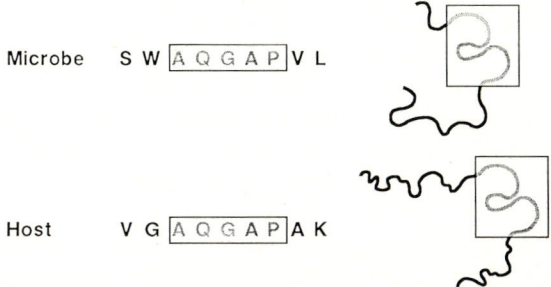

Fig. 1. Outline of the sharing of a linear amino acid sequence or of a conformational fit between a microbe and a host "self" determinant. This is the basis for the first phase of molecular mimicry. Autoimmunity can occur if (1) a host immune response raised against the sequence from the microbe cross-reacts with the host "self" sequence and (2) the host sequence is a biologic important domain, e.g., encephalitogenic site of myelin basic protein, the site on the acetylcholine receptor that is important for gaiting membrane changes needed for a synapse

---

Department of Immunology, Scripps Clinic and Research Foundation, La Jolla, CA 92037, USA

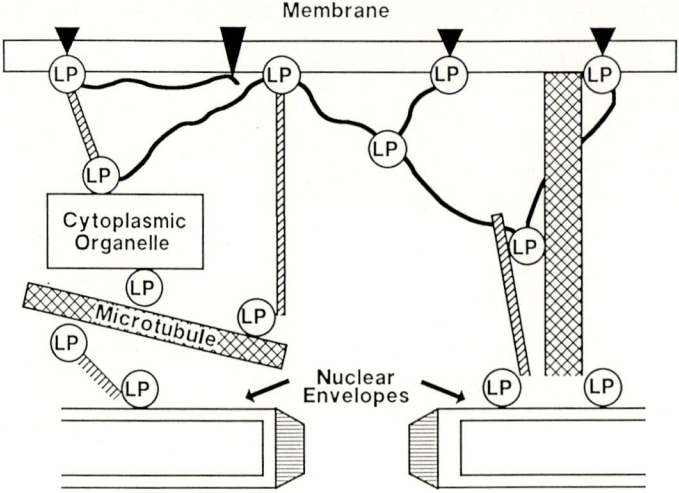

Fig. 2. Diagrammatic conceptualization of a hypothetical network linking glycopeptides at the surface with components of the CS. The linker proteins (*LP*) are of many types in complexes with microfilaments, IF, or microtubules. Actin-containing microfilaments are shown as *solid, wavy* structures and IF as the *narrower, cross-hatched* rods. At any one site, the LP may occur as a single species or aggregates of two or more species of protein. There is no attempt to represent to scale any of the macromolecular complexes and cellular organelles. (From DALES et al. (1983) CS: cytoskeletal system; If: intermediate filaments; LP: linker protein

may use the cytoskeletal apparatus for their own purposes of trafficking and assembly.

Examples of molecular mimicry were first described as such in the early 1980s by investigators who found that monoclomal antibodies against SV40 T antigens cross-reacted with host protein inside the cell (LANE and HOEFFLER 1980). However, the importance of this observation really became apparent when others realized that the monoclonal antibodies against a battery of viruses were cross-reacting with host determinants. For example, cross-reactivity between measles virus phosphoprotein (72K molecular weight) was noted with the cytoskeleton component keratin (54K molecular weight), while the herpes simplex virus glycoprotein of 140K cross-reacted with a separate epitope on keratin from that recognized by the measles virus phosphoprotein (FUJINAMI et al. 1983). Concurrently, DALES et al. (1983) found shared homology between vaccinia virus hemagglutinin and the cytoskeleton protein vimentin.

The frequency of cross-reactivity between viral proteins and host self antigens was analyzed by SRINIVASAPPA and his colleagues (1986) using over 600 monoclonal antibodies. Combining these antibodies with an additional 50 monoclonal antibodies we noted that over 4% of the over 700 monoclonal antibodies against 14 different viruses, including such commonly found representatives of DNA and RNA viruses as the herpesvirus group, vaccinia virus, myxoviruses, paramyxoviruses, arenaviruses, flaviviruses, alphaviruses, rhabdoviruses, coronaviruses, and human retroviruses, cross-reacted with host-cell determinats expressed on uninfected tissues (Table 1). Some of these monoclonal viral-specific antibodies reacted with antigens in more than one organ. These data indicated that molecular mimicry is not uncommon and not

Table 1. Molecular nimicry: virus-host antigens tested by monoclonal antibodies

Reactive with virus	No. tested	Reaction with unifected tissues
Coxsackie B4	66	1
Japanese encephalitis	34	6
Lymphocytic chroriomeningitis	174	3
Theiler's virus	64	9
Measles	39	5
Rabies	80	2
Vesicular stomatitis	37	2
Herpes simplex (I)	20	1
Vaccinia	16	1
Dengue	132	0
Cytomegalovirus		
Human	24	1
Mouse	14	0
Human immunodeficiency virus	8	1
Total	711	34 (4.7%)

Combined data from studies on molecular mimicry previously published by SPRINIVASAPPA et al. (1986) and unpublished observation. For technical details concerning the screening procedure on unifected tissues see SRINIVASAPPA et al. (1986)

Table 2. Examples of monoclonal antibodies that cross-react between a microbial protein and an interesting host self determinant

Cross reaction	Biologic implication
Coxackie B3, B4[a] Myocardium	VP1 protein of virus with the tissue involved in coxackie B virus-induced myocarditis
Measles virus[b] T lymphocyte subset	Measles virus induces immunosuppression (examples with HA and NP of measles)
Theiler's virus[c] Galactocerebroside	VP1 protein of the virus with a major structural component of the surface of oligodendrocytes, the cell surrounding the myelin sheath. Theiler's virus causes demyelination as does adoptive transfer of antibody to galactocerebroside
HIV-1[b] Glia (astrocytes)	gp41 of HIV with a glial (non-GFAP) marker. The most common pathologic finding in the brain in patients with AIDS dementia is gliosis (reactive astrocytosis)

[a] Personal communication, Abner Notkins, NIH.
[b] Unpublished data, Michael Oldstone, Scripps Clinic and Research Foundation and Robert Fujinami, UCSD.
[c] Personal communication, Robert Fujinami, UCSD.
HA: haemagglutinin; NP: nucleoprotein; GFAP: gila fibrillary acid protein

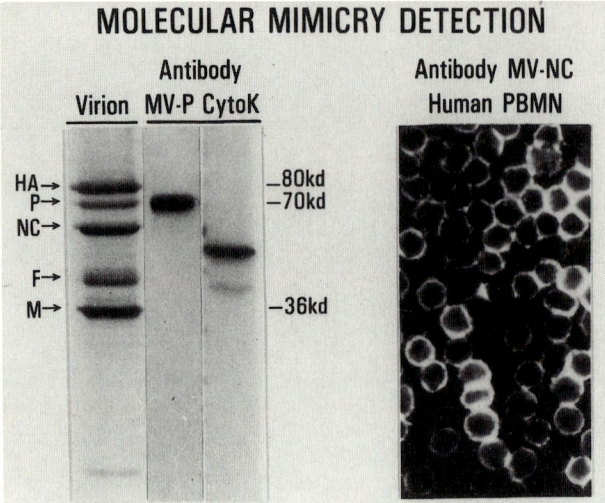

Fig. 3. *Left*, monoclonal antibody to measles virus P (*MV-P*) protein also reacts with cytokeratin (*Cytok*) (see Fujinami et al. 1983 for experimental details). *Right*, monoclonal antibody to measles virus nucleoprotein also reacts with molecule(s) on the surface of T lymphocytes (see Table 2; Oldstone and Fujinami, unpublished data)

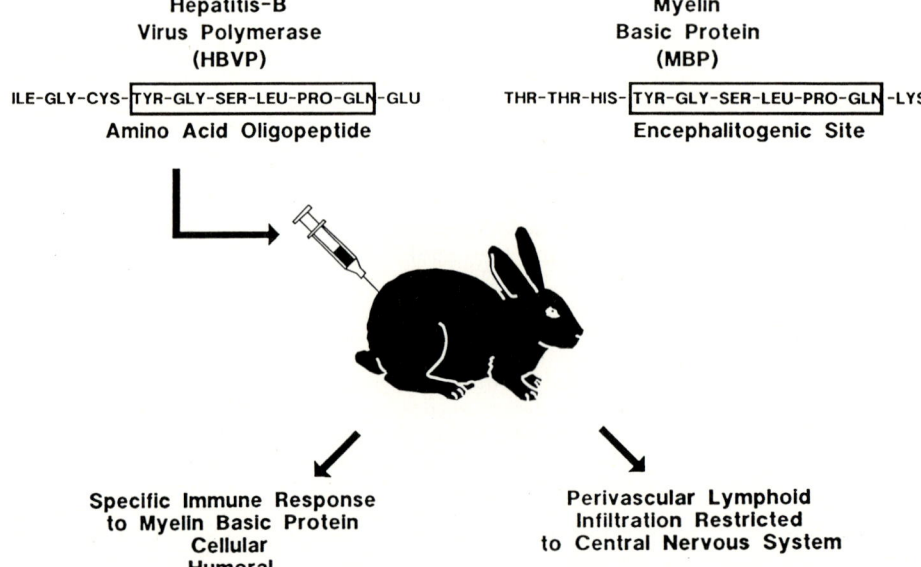

Fig. 4. First demonstration that molecular mimicry could cause disease. When New Zealand rabbits were inoculated with the 10-amino acid peptide from hepatitis B vinu polymerase (*HBVP*) they generated specific T (proliferation) and B (antibody) lymphocyte responses. Most significant, 4 out of the 11 inoculated rabbits (40%) developed histopathologic criteria for lesions of allergic encephalomyelitis (EAE). In contrast, studies with over 10 different peptides in more than 30 rabbits failed to elicit EAE

restricted to any specific class or group of virus. Most interesting, a number of these and other reported cross-reactivities (Table 2, Fig. 3) raise interesting and suggestive possibilities for the etiology and pathogenesis of a number of autoimmune diseases.

A major breakthrough for the argument that molecular mimicry could be biologically important rather than an epiphenomena resulted from experiments demonstrating that molecular mimicry could elicit autoimmune diseases. The initial observation concerned myelin basic protein and allergic encephalomyelitis (FUJINAMI and OLDSTONE 1985; Fig. 4). Myelin basic protein was chosen as the host component to study because its entire amino acid sequence is known, and its encephalitogenic site of 8–10 amino acids had been mapped in several animal species. With the use of computer-assisted analysis, several viral proteins listed in the Dayhoff files showed significant homology with the encephalitogenic site of myelin basic protein. Included were similarities and/or fits between myelin basic protein and the nucleoprotein and hemagglutinin of influenza virus, coat protein of polyomavirus, core protein of adenovirus, polyprotein of poliomyelitis virus, EC-LF2 protein of Epstein-Barr virus, hepatitis B virus polymerase, and others. However, the best fit occurred between the myelin basic protein encephalitogenic site in the rabbit and hepatitis B virus polymerase (HBVP).

Products of the immune responses, both humoral and cellular, generated in rabbits inoculated with the octamer and decamer viral peptide reacted with whole myelin basic protein. Further, inoculation of the HBVP peptide into rabbits caused perivascular infiltration localized to the central nervous system reminiscent of the disease induced by inoculation of either whole myelin basic protein or the encephalitogenic site of myelin basic protein (Fig. 4). Recently Noel ROSE and his colleagues (NEU et al. 1987) provided similar data (Fig. 5) while studying infection of coxsackie B3 virus into A/J mice. They noted that following infection, mice developed myocarditis. While part of the lesion was due to a direct viral effect, another consequence of the infection was the generation of autoantibody to myosin. Transfer of this antibody to uninfected recipients caused myocarditis. It was then found that coxsackie B3 and myosin (cardiac h chain) shared 7 out of 11 amino acids (Fig. 5). Inoculation of the shared myosin sequence resulted in myocarditis, again with the eliciting of autoantibodies to myosin.

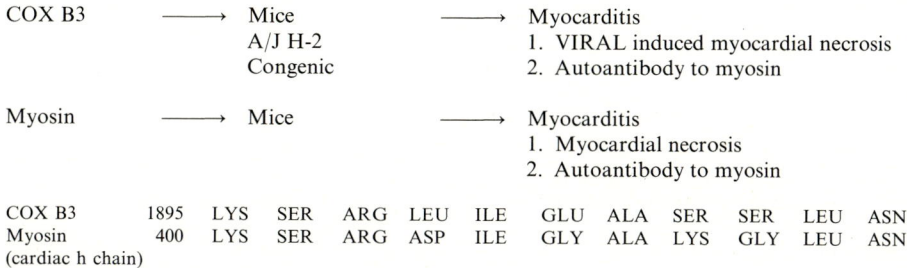

Fig. 5. Additional evidence that molecular mimicry causes autoimmune-mediated disease. Model system studied by Noel ROSE and his colleagues (see NEU et al. 1987) showing molecular mimicry between coxsackie B3 virus (COX B3) and cardiac myosin. Earlier, SRINIVASAPPA et al. (1986) demonstrated that the monoclonal antibody that neutralized coxsackie B3 also stained myocardial tissue from several mammalian species (see Table 2)

Both of the above exerpiments are seminal in that they conclusively showed that molecular mimicry could cause autoimmune responses and autoimmune disease.

The most likely explanation for how molecular mimicry caused disease is that an immune response against the determinant shared by host and virus brought forth a tissue-specific immune response, presumably capable of destroying cells and eventually the tissue. The probable mechanism was the generation of cytotoxic, cross-reactive, effector lymphocytes or antibodies that recognized specific determinants of self proteins located on target cells. Interestingly, the induction of cross-reactivity need not require a replicating agent, and the immunologically mediated injury could occur after removal of the immunogen — a hit-and-run event. Clearly, the microbial infection that initiated an autoimmune phenomenon need not be present at the time overt disease developed. A likely scenario would be that the virus responsible for inducing a cross-reacting immune response was cleared initially, but the components of that immunity continued to assault host elements. The cycle continues as the autoimmune response itself leads to tissue injury that, in turn, released more self-antigen, thereby inducing more antibodies, and so on. Such a sequence might account for the virus encephalopathies occurring in humans after measles, mumps, vaccinia, or herpes zoster virus infections; in these postinfectious diseases, recovery of the inducing agent has been rare. This theory is reinforced by studies showing that, after several types of acute viral infection, mononuclear cells from peripheral blood or cerebral spinal fluid proliferate in response to host antigens, one of which is myelin basic protein. Interestingly, several clonal populations of lymphocytes have been

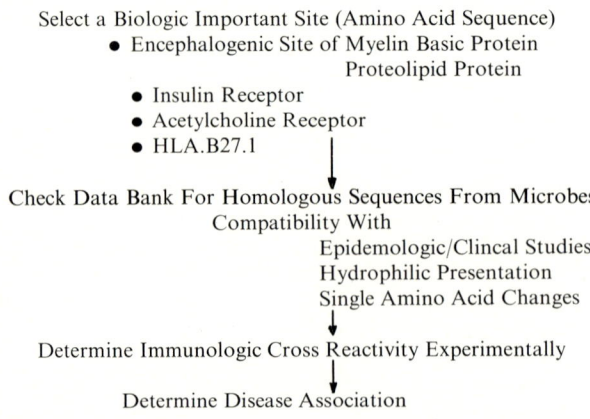

Fig. 6. Flow diagram of the tactics that can be followed to investigate human diseases suspected to have an autoimmune basis. The chapters in this volume by DYRBERG, STEFANSSON, and by SCHWIMMBECK address studies regarding the insulin receptor and viruses in diabetes, acetylcholine receptor and bacteria in myasthenia gravis, and HLA B27 (self-autoantigen) with bacteria in nonrheumatoid arthritidies, respectively. Recent observations linking virus with the acetylcholine receptor and myasthenia gravis has been documented using this approach by SCHWIMMBECK et al. (1988)

Table 3. Sequence similarities between microbial proteins and human host "self" proteins[a]

Protein	Position	Sequence
HLA-B27	70:	LYS ALA GLN THR ASP ARG GLU ASP LEU
KLEB PN	186:	SER ARG GLN THR ASP ARG GLU ASP GLU
HuAChR	160:	PRO GLU SER ASP GLN PRO ASP LEU
HSV GP D	286:	PRO ASN ALA THR GLN PRO GLU LEU
HLA DR	50:	VAL THR GLU LEU GLY ARG PRO ASP ALA GLU
HCMV IE-2	79:	PRO ASP PRO LEU GLY ARG PRO ASP GLU ASP
INSULIN r	66:	VAL TYR GLY LEU GLU SER LEU ASP LEU
PAPILLOMA E2	76:	VAL LEU HIS LEU GLU SER LEU LYS ASP SER
COAGFACT XI	269:	ILE LYS LYS SER LYS ALA LEU
DENGUE	68:	ILE LYS LYS SER LYS ALA ILE
MYOSIN	138:	TYR GLU ALA PHE VAL LYS HIS ILE MET SER VAL
COX B3	2152:	TYR GLU ALA PHE ILE ARG LYS ILE ARG SER VAL
BRAIN PROTEIN	156:	ASP SER THR LYS ASN ARG LYS THR ASP
HIV POL	222:	ASP SER THR LYS TRP ARG LYS VAL ASP

[a] Shared sequences currently under active evaluation for pathogenic role in selected diseases. Immunologic cross-reactivity has now been observed for the majority of these sequence pairs, and a disease association has been suggested for the first two pairs by investigation of specimens from patients with ankylosing spondylitis and myasthenia gravis, respectively.

harvested from central nervous system fluid of humans with encephalitis that proliferate to the infecting virus as well as to nervous system antigens.

Importantly, molecular mimicry occurs only when the virus and host determinats are sufficiently similar to induce a cross-reactive response yet different enough to break B or T cell immunologic tolerance, and the response is directed against a host "self" determinant (epitope) having an important biologic activity. Thus, an immune response against the encephalitogenic site of myelin, the effector domains of the acetylcholine or insulin receptor or the hypervariable region of HLA B27 molecule as opposed to its conserved regions are likely of importance for disease causation or restriction of disease to a specific HLA genotype.

The use of sound epidemiologic data suggesting disease association combined with interesting homogloies between the microbe in question and the biologically important epitope can be used as a strategy to explore molecular mimicry causation of disease. An outline of the approach to be followed is provided in Fig. 6. This basic outline has proven successful as related by chapters of this book for *Klebsiella pneumoniae* nitrogenase, *Shigella flexni*, *Yersinia pseudotuberculosis* with HLA B27 in ankylosing spondylitis and Reiter's syndrome; for adenovirus with alpha gliadin in celiac disease; for initial studies on diabetes looking at antibodies to the insulin receptor and papillomavirus; for herpes simplex virus (SCHWIMMBECK et al. 1988) or bacteria with the avetylcholine α receptor in myasthenia gravis; and for adjuvant arthritis myobacteria and heat shock protein. Further, this kind of approach has helped in investigating streptococcal infection and heart disease.

Recent computer searches have indicated interesting sequence homologies that might explain a variety of other diseases; for example, the amino acids shared between a number of coagulation proteins and Dengue virus or among HIV and brain proteins could suggest part of the pahtogenetic mechanism for Dengue hemorrhagic shock syndrome and AIDS dementia complex (Table 3).

In summary, molecular mimicry defines the sharing of molecular determinants between microbe and self. The frequency of occurrence is not rare and is unlikely to be due to chance. Evidence for possible associations with human diseases is accumulating although definitive proof will be difficult. Yet, the experimental animal evidence of disease causation strengthens the argument for disease in humans. Finally, at the very least, mimicry provides a series of useful probes that can be used to obtain specific cell proteins and genes. Hence, a number of laboratories are using these reagents to isolate specific sets of neurons, hormone receptors, lymphocyte subsets, and myocardial genes. In addition to providing a path to seek etiologic agents, it also provides a framework for the molecular analysis of tolerance and autoimmunity.

Acknowledgments. This is publication number 5625-IMM from the Department of Immunology, Scripps Clinic and Research Foundation, La Jolla, California 92037, USA. This work was supported in part by USPHS grants AI-07007, NS-12428, AG-04342, and U.S. Army Medical Research and Development Command Contract No. DAMD17-88-C-8103. Opinions, interpretations, conclusions, and recommendations are those of the author and are not necessarily endorsed by the U.S. Army.

References

Dales S, Fujinami RS, Oldstone MBA (1983) Serologic relatedness between Thyl. 2 and actin revealed by monoclonal antibody. J Immunol 131: 1332–1338

Fujinami RS, Oldstone MBA, Wroblewska Z, Frankel ME, Koprowski H (1983) Molecular mimicry in virus infection: cross-reaction of measles phosphoprotein or of herpes simplex virus protein with human intermediate filaments. Proc Natl Acad Sci USA 80: 2346–2350

Fujinami RS, Oldstone MBA (1985) Amino acid homology between the encephalitogenic site of myelin basic protein and virus: Mechanism for autoimunarity. Science 230: 1043–1045

Lane DP, Hoeffler WK (1980) SV40 large T shares an antigenic determinant with a cellular protein of molecular weight 68,000. Nature 288: 167–170

Neu N, Rose NR, Beisel KW, Herskowitz A, Gurri-Glass G, Craig S (1987) Cardiac myosin induces myocarditis in genetically predisposed mice. J Immunol 139: 3630–3636

Srinivasappa J, Saegusa J, Prabhakar BS, Gentry MK, Buchmeier MJ, Wiktor TJ, Koprowski H, Oldstone MBA, Notkins AL (1986) Molecular mimicry: frequency of reactivity of monoclonal activiral antibodies with normal tissues. J Virol 57: 397–401

Schwimmbeck PL, Dyrberg T, Drachman D, Oldstone MBA (1988) Molecular mimicry and myasthenia gravis: uncovering a novel site of the acetylcholine receptor that has biologic activity and reacts immunochemically with herpes simplex virus. J Clin Invest (manuscript submitted)

Subject Index

acetylcholine 3
—, receptor 2
—, α receptor 134
AchR, α-chain 57
acid, hyaluronic 7
actin 104
activator, polyclonal 2
adenovirus 2 68, 134
— 12, Ad12-neutralizing antibody 73
——, prior exposure 73
— 18, 73
— E1b protein and A-gliadin, sequence similarity 70
— proteins, intestinal 70
agents, infectious 1
AIDS dementia complex 134
amino acid sequence homology 70
ankylosing spondylitis 2, 45, 134
antibodies 132
—, cross-reacting 7
—, —, diseases associated with 6
—, DNA-specific 1
—, monoclonal 14, 61
—, nonprecipitating 8
antibody, contaminating 104
— cross-reactivity 6, 60
—, eclipsed 101
—, gliadin-specific 76
—, merozoite surface 105
—, titers, neutralizing 75
—, ring-infected surface 110
antigen-binding site 62
—, —, host self 128
antigens, histocompatibility 16
—, HLA 76
—, membrane 13
—, —, components 13
—, —, heart reactive antibodies and 14
—, —, monoclonal antibodies to 14
—, —, tolerazing phenomena 17
—, transplantation 15
arthritis 2
—, adjuvant 30

—, adjuvant-myobacteria 134
Aschoff lesion 20
——, subpopulation of lymphocytes in 20
autoantibodies 58, 118
—, cytoskeletal 127
autoantigens 117
autoimmune disease 1, 60, 129
— mechanisms 17, 82, 83
——, idiotype networks 82
——, suppressor T-cells 82
— response 58
autoimmunity 93, 98, 101, 102, 110, 117, 127
—, virus-induced 92, 99

BiP, see immunoglobulin heavy chain binding protein
blood group 102, 104, 108, 109
Brugia malayi 103

camouflage hypothesis 106
carbohydrates, group A 12
cartilage 31
celiac disease 67, 134
——, dietery grains that activate 68
——, etiology 68
——, pathogenesis 68
cells, B 21
—, β 117
—, cytotoxic 76
central nervous system (CNS) 93, 94, 96, 97, 98
———, disease 93, 96
Chagas' disease 79—92
——, denervation 81
——, genetic basis 89
——, indeterminate phase 80, 82
——, latent phase (see indeterminate phase)
——, megasyndromes 81
——, pathology, acute 80
——, —, cardiac 80, 81, 84, 85
——, —, chronic 81—82
——, vaccination 87
coagulation factor XI 133

Subject Index

compliment, binding   106, 110
—, damage   104, 106
computer search   48
conservation, molecular   103, 104
coxsackie virus   12
— —, B   133
— —, B3   131
— —, B3 and B4 infection relation to auto-
    immune myocarditis   12
cross-immunity, nonreciprocal   103
cross-immunization   103
cross-reacting antibodies, associated
    diseases   6
cross-reaction   132
— —, at level of T-cell recognition   76
— —, immunologic   75
CSP, see protein, circumsporozoite
cytoskeleton   106

Dayhoff Data Base   48
Dengue hemorrhagic shock syndrome   134
— virus   133, 134
dermatitis herpetiformis   76
determinants, cross-reacting   58
diabetes   2, 134
—, insulin-dependent   117
—, insulin-resistant   117
dietery grains that activate celiac disease   68
disease, demyelinating   2
—, postinfectious   132
DNA viruses   1

EAE (see experimental allergic encephalo-
    myelitis)   94, 95, 96
echovirus 11   73
EGF (see epidermal growth factor)
encephalomyelitis, allergic   131
—, —, experimental   94
epidermal growth factor   105, 106
epitope, aggressive   102, 106
—, —, mimetic   102
—, autoimmunogenic   102, 103, 106, 110
—, carbohydrate, see oligosaccharide epitope
—, eclipsed   101, 109–110
—, —, mimetic   101
—, functional   105
—, oligosaccharide   107, 109
—, —, N-linked, complex type   107
—, —, —, high mannose type   107
—, —, O-linked   107
—, —, mimicry   105
epitopes, bacteria   60
—, shared   60
epitasis   108–109
Epstein-Barr (EB) virus   34
EVI immunofluorescence   80, 81, 83, 86
expression, enhanced, MHC class I + II   2

Fasciola hepatica   109
fibronectin (FN)   106, 110
filarid   103
flexibility   60

galactocerebroside   93, 96
gene bank   14
—, codominant   108
—, cloned   104
—, immune response   37, 108
—, — suppression   108
gliadin   7
—, A-gliadin   69
—, —, heptapeptide   72
—, α   2, 69, 134
—, — sequences   70
—, β   69
—, challenge   70
—, γ   69
—, — sequences   70
—, ω   69
glomerulonephritis   60
—, post-streptococcal   18
—, —, in monkeys   18
—, —, genetics   19
—, —, and M-protein   19
glutamile   71
glutenin   67
glycolipid   107
glycoprotein   105, 107
— D   59
glycosyl transferase   109
glycosylation   36
group A carbohydrate, chemistry   12
— —, reaction with heart valve glyco-
    protein   13
— —, serology   13
— streptococcus, skingraft rejection and   16
gut flora   61

H-2   108
HIV   134
Heligmosomoides polygyrus, see *Nematospiroi-
    des dubius*
heredity   17
—, role in rheumatic fever   17
herpes simplex virus   59, 133, 134
histocompatability antigens and M-protein   16
— — and post-streptococcal glomerulone-
    phritis   19
— — and rheumatic fever   18
hit-and-run event   132
HLA   108
— B27   2, 45, 133, 134
— DR   133
— class IID region   67
— haplotypes   29

— phenotypes 58
homograft rejection 15
hsp, *see* protein, heat-shock
human adenovirus serotype 12, Elb protein 70
— cytomegalovirus 133
— immunodeficiency virus 133
hyaluronic acid, biochemistry 7
— —, non-precipitating antibodies, in man 8
— —, — —, in rabbits 8
— —, serology in animal and man 8
hydrophilicity 49, 60
hydrophobic region 60

idiotypes 40, 62, 104
immune response 67
— —, genes 46
— —, cell mediated 67
— —, genes 46
— —, humoral 67
immunity, cell-mediated 20
—, evasion of 101, 102, 103, 109, 110
—, exploitation of 108
immunogenetics 58
immunoglobulin 103, 110
immunopathogenesis 108
inflammation 103
insulin-dependent diabetes (IDDM) 117
— receptor 117, 133
— resistant diabetes 117
Ir gene, *see* gene, immune response
Is gene, *see* gene, immune suppression

Klebsiella pneumoniae 2, 133, 134
— — nitrogenase 48

lectinophagocytosis 107
Leishmania braziliensis 108
— *donovani* 108—109
leishmanial 105, 107
leishmaniasis 108—109
leprosy 28
lipid, bilayer 106
lymphocytes 132
— B 2
— —, antigen in patients and relatives with rheumatic fever 21
— T 2
lymphocytic choriomeningitis virus 1
lymphokines 1, 2

macrophage 105, 107
major histocompatibility locus complex 67
malaria, *see Plasmodium*
marker 18
masquerade 110
—, antigenic 110
MBP (*see* myelin basic protein)

merozoite 106
MHC 28
microorganism 61
mimicry 32
—, antigenic 102
—, biological 103
—, molecular 2, 5, 47—50, 61, 76, 117, 127
—, —, definition 7
—, —, origin of term 5
—, oligosaccharide 105
—, peptide 105
—, receptor/ligand 105, 106, 109
modifying factor 46
monoclonal antibodies 14, 61
monokines 2
mosquito 106
M protein 8
mucosa, intestinal 76
—, small intestinal 67
myasthenia gravis (MG) 2, 57, 122, 134
Mycobacterium bovis BCG 33
— *leprae* 28
myelin basic protein 93, 94, 97, 98, 99, 131
myobacteria 2
—, adjuvant arthritis 134
myocardial tissue 2
myocarditis 131
myosin 11, 133
—, antibodies in myocarditis 11—12

Nematospiroides dubius 108
network, idiotypic 59
New Zealand mouse 1
nicotinic acetylcholine receptor (AchR) 57

ookinete 106

papillomavirus 120, 134
parasitism, intracellular 105, 106, 110
paratope 62
parvovirus H1.VP2 protein 59
pathogenesis 117
Pch 105, *see* antigen, ring-infected surface
peptide hydrolases 68
—, mimicry 105
—, sequence homology 102, 105
—, synthetic 33, 104
Pf 155, *see* antigen, ring-infected surface
PFs 25 105, 106
Plasmodium 104, 105, 106, 107
platelet 105
poliovirus 121
polyomavirus middle T antigen 59
polyclonal activator 2
polymorphism, antigenic 101, 102, 110
—, immune suppression gene 108, 110
postinfectious encephalomyelitis,

pathogenesis 59
precipitates, acetone 35
prolamins 68
proline 71
properdin 107
protein, actin cross-linking 104
—, Ad12 E1b 76
—, 54K alb 71
—, brain 134
—, circumsporozoite 105, 107
—, coagulation 134
—, erythrocyte band 3 105, 106
—, heat-shock 2, 32, 103, 134
—, histocompatibility 109
—, immunoglobulin heavy chain binding 103
—, intermediate filament 105
—, intestinal adenovirus 70
—, link 34
—, M 8
—, —, contribution to virulence 8
—, —, significance of membrane anchor 13
—, —, streptococcal 2
—, —, structure 10
—, thrombospondin-related anonymous 105, 107
— thyroid 3
proteoglycan 32
—, cartilage 2
—, role in glomerulonephritis 10
— turnover 36

receptor, acetylcholine 134
—, C3 106
—, CR 1 107
—, complement 106, 107
—, Fc 101, 106
—, insulin 3, 134
—, mannose-fucose 107
Reiter's syndrome 2, 45, 60, 134
RESA, see antigen, ring-infected surface residues, synthetic M 5 10
response, autoimmune 58
restriction fragment length polymorphism, 4-kb RsaI DPβ-chain genomic fragment 76
RGD, see sequence, RGD
rheumatic fever 17, 60
——, autoimmune role in pathogenesis 17
——, B-cell antigen 17
——, cell mediated immunity and 20
——, heredity 17
——, MHC association and 18
——, ——, possible marker 18
——, ——, presence in families 18
——, ——, significance in pathogenesis 21
RNA virus 1

schistosome 101, 103, 106, 107, 108, 109

self-tolerance 38, 109
sequence, homology 102, 105—107, 109
—, —, amino acid 70
—, RGD 106
— similarity between A-gliadin and the adenovirus 12 E1b protein 70
sera, Ad5 E1b-specific 72
Shigella flexni 134
snail 107
spine, schistosome tegumental 104
spondylitis ankylopoietica 60
streptococcus, group A 7
suppressor cell, I—E restricted 108
— inducer cell 39
Sydenham's chorea 14
synovial fluid 36
— tissue 51
synthetic peptides 49
——, M-protein, Ile-Arg-Leu-Arg the "glomerular" epitope 11
——, —, pep 5M, cross-reaction with sarcolemmal antigen 10
——, —, ——, relation to opsonization 10—11
——, —, peptide 164—197 11

T cell 102, 110
—— epitope 2, 33
thrombospondin (TSP) 105
thymona 58
thyroid disease 122
— microsomal antigen 122
tolerance to self 28
transplantation antigens, epitope sharing with streptococcal antigens 15
——, homograft rejection and 15
——, renal transplant, strep infection and 15
TRAP, see protein, thrombospondin-related anonymous
Trichinella spiralis 108
trichomonad 106
trigger, developmental 103
Trypanosoma cruzi 2, 79—92
——, antigens 83—86
——, —, adsorption onto host cells 83, 88
——, —, expression on host cells 88
——, —, idiotype-specific antibodies 87, 88
——, —, MHC affinity 88
——, —, multigene families 84
——, —, recombinant DNA 84, 85
——, antigenic variation 84
—— cross reactivity 83, 85—88
————, carbohydrate epitopes 87
————, connective tissue 86
————, cruzin 85
————, fibronectin receptor 85
————, heat shock proteins 85

––––, laminin 83, 86, 87
––––, monoclonal antibodies 87
––––, muscle 83
––––, neurons 83
––––, schwann cells 83
––––, 67K protein 85
––, detection of parasites 80
––, insect vector 79
––, life cycle 70, 80
– *rangeli* 87
trypanosomatid 103, 106
twins, monozygotic 70

vaccine 106, 107
vacuole, parasitophorous 106
virus 102, 105, 106

–, coxsackie 12
–, Dengue 133, 134
–, DNA 1
–, encephalopathies 132
–, human immunodeficiency 133
–, lymphocytic choriomeningitis 1
–, papilloma 120, 133
–, polynoma 1
– RNA 1

Western blots 61
wheat, durum 69
–, endosperm 69
– gluten 67

Yersinia pseudotuberculosis 134

OCT 16 1989